THE GENERAL PRINCIPLES FOR THE RECKONING OF TIME IN CANON LAW

THE CATHOLIC UNIVERSITY OF AMERICA
CANON LAW STUDIES
No. 144

THE GENERAL PRINCIPLES FOR THE RECKONING OF TIME IN CANON LAW

AN HISTORICAL SYNOPSIS AND COMMENTARY

A DISSERTATION

Submitted to the Faculty of Canon Law of the Catholic University of America in Partial Fulfillment of the Requirements for the Degree of

DOCTOR OF CANON LAW

BY

ARTHUR JOSEPH DUBÉ, A.B., J.C.L.
Priest of the Diocese of Portland

THE CATHOLIC UNIVERSITY OF AMERICA PRESS
WASHINGTON, D. C.
1941

NIHIL OBSTAT:

CLEMENT V. BASTNAGEL, J.U.D.,
Censor Deputatus.

IMPRIMATUR:

✠ JOSEPH EDWARD McCARTHY,
Episcopus Portlandensis.

Portland, May 17, 1941

PRINTED IN THE UNITED STATES OF AMERICA
BY THE WATKINS PRINTING CO., BALTIMORE

TO

MY PARENTS

IN

GRATEFUL ACKNOWLEDGMENT

TABLE OF CONTENTS

TABLE OF CONTENTS v

FOREWORD ix

INTRODUCTION 1

I—Definition of Time 1
II—Importance of Time 2

PART I

The Material Reckoning of Time 5

CHAPTER I. THE CALENDAR 7
Article I. The Various Forms of Ancient Calendars 7
Article II. The Evolution of our own Calendar 11
§ 1 Before Caesar 11
§ 2 Julian Reform 15
§ 3 Augustan Changes 16
§ 4 Gregorian Reform 17
Article III. Accidental Details 23
§ 1 The Beginning of the Year 23
§ 2 The Meaning of "1941" 25
§ 3 Papal Documents 30
Article IV. The Component Parts of the Calendar 30
§ 1 The Year 31
§ 2 The Month 33
§ 3 The Week 35

CHAPTER II. THE CLOCK 38
Article I. The Primitive Concept of Day 38
Article II. The Modern Concept of Day 41

PART II

The Juridical Reckoning of Time 51

SECTION I

Historical Synopsis 51

CHAPTER III PRELIMINARY OBSERVATIONS 51
Article I. General Remarks 51
Article II. The Jewish Reckoning 52
CHAPTER IV. ROMAN LAW 58
Article I. The Measuring of Time 58
Article II. The Reckoning of Time 59
§ 1 General Remarks 59
§ 2 The Natural Reckoning 61
§ 3 The Civil Reckoning in General 61
§ 4 The Question of the *Terminus a Quo* and the *Terminus ad Quem* 63
§ 5 The *Dies Intercalaris* 65
§ 6 Special Reckoning of the Year and the Month 66
Article III. *Tempus Utile* 66

CHAPTER V. CANON LAW 71
Article I. Preliminary Remarks 71
Article II. The Duration of the Year and the Month 73
§ 1 The Year 73
§ 2 The Month 76
Article III. The Reckoning of Time 78
§ 1 The Natural Reckoning 79
§ 2 The Problem of the Last Year or Month 80
§ 3 The *Dies Terminus ad Quem* 84
§ 4 The *Dies Terminus a Quo* 87
Article IV. Physical or Moral Reckoning 87
Article V. The Day and the Hours 89
Article VI. Dawn and Dusk 101
Article VII. A Few Additional Remarks 103
§ 1 *Tempus Utile* 103
§ 2 The *Dies Certa* and Similar Expressions 104
§ 3 The Expressions *Statim* and *Quamprimum* 104
Article VIII. The Immediate Sources of the Code 105

SECTION II

Practical Commentary of Canons 31-35 108

CHAPTER VI. UNDERLYING PRINCIPLES 108

Article I. Introductory Remarks 108
Article II. The Fundamental Norm 111
§ 1 Liturgical Laws 111
§ 2 Explicit Exceptions 114
§ 3 Contracts 116
§ 4 Implicit Exceptions 121
§ 5 Extension of the General Norms 123
Article III. The Duration of Time 125
§ 1 The Day 125
§ 2 The Week 129
§ 3 The Month and the Year 130

CHAPTER VII THE RECKONING OF THE HOURS 131
Article I. The Institutes Involved 131
§ 1 The Private Celebration of Mass 131
§ 2 The Private Recitation of the Breviary 135
§ 3 The Reception of Holy Communion 136
§ 4 The Observance of the Law of Fast and Abstinence 138
Article II. Exhaustive or Illustrative Enumeration 138
Article III. The Use of the Various Reckonings 142
§ 1 The Various Reckonings Permitted by Canon 33, § 1 142
§ 2 The Double Probability 153
§ 3 When Two or More Reckonings are Involved 158
A. The Various Opinions 159
B. The Solution of the Problem 163
§ 4 Other Problems 169
A. The Case of the Same Obligation for Two or More Successive Days 169
B. The Actual Determination of a Reckoning 175
§ 5 The Reckoning of the Day for Travelers 180
A. When Traveling from One Zone to Another 180
B. When Crossing the International Date Line 186

CHAPTER VIII. THE RECKONING OF THE LARGER UNITS OF TIME 189
Article I. Introductory Remarks 189
Article II. The Reckoning According to the Calendar.... 189

§ 1 When is the Calendar Reckoning to be Used? 189
§ 2 Which Calendar is to be Used? 191
A. The Problem of the Leap Month or Year 192
B. The Other Calendars 195
Article III. The Natural and the Civil Reckoning 200
§ 1 The Question of the Determination of the Starting Point 200
§ 2 The Natural Reckoning 210
§ 3 The Civil Reckoning 217
A. Determination of the First Juridical Day 218
B. Reckoning when a Day of Identical Date is Lacking 221
C. The Repetition of Acts of the Same Nature 224

CHAPTER IX. OTHER CONSIDERATIONS 230
Article I. Available Time 230
§ 1 Introductory Remarks 230
§ 2 The Reckoning of Available Time 233
§ 3 The Occurrence of Available Time 240
Article II. The Moral Reckoning of Time 244
§ 1 *Parum pro Nihilo Reputatur* 245
§ 2 *Dies Incepta Habetur pro Completa* 251
§ 3 The Completion of Time 252

CONCLUSIONS 255

APPENDIX I 257

APPENDIX II 265

BIBLIOGRAPHY 273

LIST OF ABBREVIATIONS 281

BIOGRAPHICAL NOTE 283

ALPHABETICAL INDEX 285

CANON LAW STUDIES.................................... 293

FOREWORD

The aim of the writer in this present work is to comment on canons 31-35 of the Code and to expound the general principles for the reckoning of time in Canon Law. It has not been found possible to include within the scope of this dissertation considerations on the many canons throughout the Code in which the computation of time is involved. The lengthy preamble to the commentary proper, it is hoped, will serve a useful purpose in supplying information necessary to the correct interpretation of Title III of Book I.

Many of the problems affecting the reckoning of time in general have been considered and a practical answer has been given in each case. A few of the solutions proposed might be considered new or daring. They are by no means considered as final. It is hoped, on the contrary, that they will invite canonists to a thorough and conscientious examination of the problem of time-reckoning.

The writer avails himself of this opportunity to express his gratitude to His Excellency, the Most Rev. Joseph E. McCarthy, D.D., Bishop of Portland, for the opportunity of advanced study, and to thank most sincerely all those who have in some way or other helped him in the preparation of this dissertation. He wishes especially to acknowledge his indebtedness to the members of the Faculty of the School of Canon Law for their kind and generous direction.

INTRODUCTION

"What then is time? If no one asks me, I know: if I wish to explain it to one that asketh, I know not." [1]

I—Definition of Time

Time is one of the intriguing entities that we constantly deal with but that we cannot easily define. Past generations have left us numberless definitions of time, but none perhaps is more worthy of mention than that given by Aristotle and adopted by St. Thomas: "Numerus seu mensura motus secundum prius et posterius." [2] Time therefore is: "The reckoning of motion as previous and subsequent." [3]

The nature of time is more easily grasped when this form of duration is compared with eternity and aevum. There are indeed three kinds of duration: "Eternity, aevum and time; the first being the duration of a thing which is altogether unchangeable, the second that of a thing which is subject to accidental change, though it remains immutable in its substance, the third the duration of a thing which is subject both to substantial and accidental change." [4]

In our minds time is closely associated with the physical changes that occur in the world around us. Whether or not these events have any relation of cause and effect, we instinctively group a certain number of them as having happened simultaneously with, or before, or after another event. Time in its more absolute acceptation simply means the complete flow of "befores" and "afters" that began with the creation and that are to endure

[1] St. Augustine, *Confessions,* Translated by E. B. Pusey (New York: Dutton, 1939), lib. XI, c. XIV.

[2] St. Thomas, *Opera Omnia,* Leonine Edition, Vol. II (Romae, 1884), *Commentaria in Octo Libros Physicorum Aristotelis,* lib. IV, lect. 17.

[3] Cicognani, *Canon Law,* Authorized English Version by O'Hara and Brennan (Philadelphia: Dolphin Press, 1935), p. 664.

[4] Phillips, *Modern Thomistic Philosophy,* Vol. I (2. ed., London: Burns, Oates and Washbourne, 1939), p. 118.

as long as matter is subject to substantial changes. Some of nature's phenomena take place at regular intervals; such are the recurrence of light and dark, the moon's phases and the perpetual cycle of the seasons. These are the units that nature has given us and it is according to them that we measure time. Any length of time is meaningless if we cannot express it in terms of years, months, weeks, days, hours, or some other unit of time-measurement.

Time is either past, present, or future, but in reality the past and the future alone have any mathematical value. Time is a continuous flow and the present is only an imaginary point dividing the past from the future. A fast moving projectile might take a few seconds to reach any given point along its murderous path and it might keep on going for seconds afterwards, but the moment at which it passes that point cannot easily be expressed in time-units. The present really has a meaning only when it is applied to some measure such as a day, or a year, etc. We call the current duration of that established measure "present time." For example, this year, or this week, or this day may be designated by the phrase "present time." [5]

II—Importance of Time

Time is and always has been a universal factor of no mean importance. Truly, time does not enter as a constituent part in the making of any one thing. Yet nothing here on earth can escape it, can be thought of independently of it. "All things have their season, and in their times all things pass under heaven. A time to be born and a time to die." [6]

It may be safe to assert that in no other epoch than the present one has time played so vital a part. Time-tables are everywhere to be found; time-locks, time-bombs, etc., are but a few of the many modern gadgets whose proper functioning depends on the element of time. The modern city is enthralled at the feet of time. The utmost accuracy is required in the time-tables that

[5] Cf. Cicognani, *Canon Law,* p. 665.

[6] Ecclesiastes, III, 1-2.

regulate the movement of trains in and out of a busy station. A simple miscalculation might have the most tragic effects.

The importance of time is not relegated to civil affairs. This element also plays a prominent part in the legislation of the Church, whether we consider its liturgical or its disciplinary laws. The fixing of Easter, for example, which regulates the liturgy of the movable feasts throughout the greater part of the year is a chronometrical problem.

If we now consider the disciplinary laws with which this dissertation is more particularly concerned, the value of time-reckoning becomes more apparent. Time is nowhere the efficient cause of our rights, but it often is the medium through which rights are acquired or lost. The canons in which time-reckoning is involved number well over 300. In many cases the validity of an act is at stake. If, for instance, the age prescribed for admission to the novitiate,[7] for the religious profession,[8] for Matrimony,[9] has not been reached according to the reckoning imposed by canons 31-35, these acts are *ipso facto* null and void. The same conclusion also applies to the reckoning of the year of the novitiate.[10]

In addition to the general considerations, a practical knowledge of canons 31-35 is a necessity to priests and even to many laymen. The daily recitation of the breviary and the Eucharistic fast are considered in canon 33, § 1. Generally, it is true, midnight slips by during one's sleep. Confessions, however, sick calls and other forms of parochial work often impose vigils on the priest. The proper computation of midnight then becomes a practical problem to be reckoned with.

Title III of Book I of the Code abounds in technical expressions that are not purely canonical. Thus one reads such words as *calendar, local time, true* or *mean, legal time, regional* or *extraordinary*. These terms are not creations of jurisprudence.

[7] Canon 542, n. 1. Cf. also canon 555, § 1, n. 1.

[8] Canon 572, § 1, n. 1. Cf. canon 573.

[9] Canon 1067, § 1.

[10] Canon 555, § 1, n. 2.

They have a meaning of their own independently of their use in Canon Law, which simply borrowed them from astronomy, chronology, or every-day use. It is therefore highly useful to examine these words in their original meaning before attempting to consider them with their intricate juridical applications. That accounts for the division of this dissertation into two parts:

Part I—The Material Reckoning of Time.
Part II—The Juridical Reckoning of Time.

PART I

THE MATERIAL RECKONING OF TIME

Time as we know it today is measured by two instruments, the clock and the calendar. The former tells us the hours of the day, the latter indicates the day, the week, the month and the year in which we are. The day and the year are in a class by themselves, the week and the month being nothing more than simple subdivisions of the year, just as seconds, minutes and hours are subdivisions in relation to the day. We now have 60 seconds to the minute, 60 minutes to the hour, 24 hours to the day. The week consists of 7 days and the month has either 28, 29, 30 or 31 days. This arrangement does not result from any intrinsic necessity; an arrangement of a different character could be accepted just as well. As a matter of fact the week and the month have had various values throughout the ages. If conventions among men and nations would call for it, a minute could just as readily be divided into 100 seconds. All these units are conventional, that is, man-made, just as are our various units of length, weight and volume. We have our inches and feet, our ounces and pounds, our pints and gallons. In other nations a simplified metric system gives rise to other standards of length, of weight, and of volume.[1]

The day and the year are to be viewed from another angle. The calendar and the clock in indicating these units are in some way handy reproductions of the celestial clock and have value only insofar as they keep the same time. Clocks and watches will sometimes run fast or slow. They must then be set. And so it

[1] The French Republican calendar used the decimal system. The day, reckoned from midnight to midnight, was divided into 10 parts or hours each containing 100 minutes. There were 100 seconds to the minute. The week was abolished and a period of ten days was adopted in its stead. These were prosaically called: Primidi, Duodi, Tridi, Quartidi, Quintidi, Sextidi, Septidi, Octidi, Novidi. Decadi.—Cf. Cappelli, *Cronologia, Cronografia e Calendario Perpetuo* (2. ed., Milano: Hoepli, 1930), p. 154. This work will hereafter be referred to as *Cronologia.*

is with calendars. The calendar has been known to lag behind the sun. It was adjusted or regulated, for instance, by Julius Caesar, by Augustus, by Gregory XIII. They set the calendar just as we set our watches. After all, the difference between calendars and watches is merely accidental; some day our calendars might be mechanical just as our clocks, especially if the 13 month calendar is ever adopted.

The celestial clock, as we look upon it today, is constituted by the double motion of the earth. This planet rotates on its axis and revolves around the sun, the former motion giving us the day, and the latter the year. Our modern instruments have found the exact duration of these units. The day is made up of 24 hours, and the year of 365 days, 5 hours, 48 minutes and 46 seconds.[2] That knowledge has made feasible the Gregorian calendar, which is a practically perfect time-keeper. In former years, however, the exact duration of the earth's revolution around the sun was an unknown quantity. Man's knowledge of time-units was not determined by cyclical calculation, but by direct observation, which was dependent on ideal weather conditions. The same phenomenon, the new moon for instance, was ofen seen on different days in various localities.

[2] Cf. Flammarion, *Astronomie Populaire* (Paris, 1890), p. 27.

Chapter I

THE CALENDAR[1]

Article 1—The Various Forms of Ancient Calendars

Besides the occurrence of light and dark the most easily observed phenomenon was the regular succession of the moon's phases. It was soon found that the month or lunation was but a short cycle within a longer one. The earth itself was the scene of various phenomena recurring at nearly regular intervals. There was a definite cycle in nature. Flowers began to bloom at at certain time, later on they withered. Egypt was favored at a given point of time with the famous Nile Flood. Along the seaboard, people noticed that the tides followed a definite cycle. In some parts of the world winds of a definite character blew at regular intervals. According to Wilson: "Over much of Southern Asia the prevailing wind blows from the southwest from May to November, and northeast from November to May." [2]

In the course of time these various seasonal changes were attributed to their proper cause, the obliquity of the ecliptic; from then on the year ceased to be measured according to the above mentioned variable seasonal changes, but was calculated by the mathematical recurrence of the equinoxes and the solstices.

The problem that faced the calendar builders of antiquity was to blend together these two units which have no common divisor. A solar year is 365¼ days long, while the month contains about 29½ days. Twelve months give us 354 days, while thirteen months mount up to 383½ days. The problem could be solved three different ways. Man could either use the one or the other of these units separately or try to combine the two.

a—Lunar calendar—In this calendar the determining unit is the month. The length of the year is arbitrary. The months

[1] From the Latin "Kalendae," the first day of the month.

[2] *The Romance of the Calendar* (New York: Norton, 1937), p. 31.

follow one another in regular succession, and their beginning is actually determined by the phases of the moon. This form of calendar is undoubtedly one of the oldest. The year is composed of 12 months of 29 or 30 days in alternating sequence. It is evident that these months have no fixed standing as to the seasons. The first month for instance will fall in winter for a while, then in autumn, etc.

The Mahometan calendar is of this type.[3]

b—Solar calendar—In this case the year is based entirely on the motion of the earth around the sun. The moon's motion is neglected, and the month is given but an arbitrary length. The year here mentioned is the tropical year of 365 days, 5 hours, 48 minutes, 46 seconds, of which there remains more to be said shortly. Our own Gregorian calendar is a typical example of the solar calendar.

The Egyptians were among the first to use the solar calendar. Their year was reckoned to have 365 days. This period was divided into 12 nominal months of 30 days, which gave 360 days. Five epagomenal or "extra-monthly" days were added to make up the required 365. These were intercalated holidays, the birthdays of five gods.[4]

This calendar had three seasons of four months; winter, spring, summer. It is still used by the Coptic community (Monophysite persuasion), and also in Ethiopia.[5] Persia, in about 500 B. C., also adopted this calendar, which was handed down to Armenia where, in a modified form, it is still in use.[6]

The beginning of the year coincided with the heliacal rising of Sirius[7] and the inundation of the Nile. There was no leap year and the difference of nearly six hours was soon felt. Spring

[3] Cf. Cappelli, *Cronologia*, p. 22.

[4] Cf. Wilson, *The Romance of the Calendar*, pp. 68-82.

[5] Cf. Wilson, *op. cit.*, pp. 70-71.

[6] Cf. Wilson, *op. cit.*, pp. 83-84.

[7] Sirius or the Dog Star (Sothis in Egyptian) is the brightest star in the heavens and has been observed for many centuries in relation with astronomical computations. The heliacal rising of Sirius simply means the moment when this star is first seen in the heavens with the rising sun.

was gradually advancing into summer, then into winter, and back again into its proper place. This cycle was completed in 1461 years and was called the Sothic cycle from the fact that the first day of the Egyptian calendar coincided with the heliacal rising of Sirius: 1461 Egyptian years of 365 days equalled 1460 Sothic years of 365¼ days. According to Censorinus (3rd. century A. D.) one of these Sothic cycles began in 139 A. D. These cycles are important for the reckoning of ancient dates.[8]

Ptolemy III († 221 B. C.) proposed a plan for the adoption of a leap year, but in vain. It was only two centuries later that every fourth year was given 366 days.[9]

c—Luni-solar calendar—This form is a combination of the preceding varieties. An effort was made to keep the true length of the month and of the year, an impossible feat. The year, as has already been shown, is not an exact multiple of the month; 12 months mount up to 354 days, while 13 months give us 383½ days. The solar year has 365¼ days approximately. These luni-solar calendars, nominally of 12 months, were adjusted periodically by the intercalation of an additional or 13th month.

The Babylonian, Jewish, Greek and Old Roman calendars were luni-solar. In many cases cycles were used to find a common multiple of the year and the month. The most noteworthy, if not the most satisfactory, of these cycles was the one invented by Meton in or about 432 B. C. He discovered that 19 solar years contained approximately 235 lunations:

235 lunations—6939 days 16 hours 31 minutes
19 Julian years—6939 days 18 hours 0 minutes
19 solar years—6939 days 14 hours 27 minutes.[10]

The difference between 235 lunations and 19 true solar years amounted to about two hours. Twelve regular months a year gave 12 x 19 or 228 months. To fill the gap Meton intercalated 7 em-

[8] Cf. Wilson, *The Romance of the Calendar*, pp. 76-77.

[9] Cf. Wilson, *op. cit.*, p. 78.

[10] Cf. Godfray, *A Treatise on Astronomy* (4. ed., London, 1886), pp. 292-293; Newcomb and Holden, *Astronomy* (3. ed., New York, 1887), p. 134.

bolismic months in his cycle. According to Petavius (17 century A. D.) and most of the authorities these were introduced one by one in the 3rd, 6th, 8th, 11th, 14th, 17th and 19th year.[11]

According to Newcomb and Holden[12] the Athenian authorities ordered the division and numbering of the years in the new calendar to be inscribed on public monuments in letters of gold. That is the origin of the Golden Number, which is nothing more than the particular number of any given year within the metonic cycle. The year 1 B. C. begins the cycle now in use. The rule to find the Golden Number of any year is to add 1 to the date and divide by 19. The remainder is the Golden Number. If the remainder is 0, the number is then 19. Thus the Golden Number for 1939 is 2: 1939 + 1 = 1940 ÷ 19 = 102 times plus 2.[13]

This number is of no value in the disciplinary reckoning of time in the Church. It is however employed in liturgical computations.

The Greek month was divided into 3 decades of 10 or 9 days each.[14]

One reads in Wilson:

> The deeper we plunge into the past, the more numerous are the calendars that we find to have been in active operation. Not only did many civilizations—Mayan and Scandinavian, for instance — have calendars distinctive of themselves. In Babylon, 4500 years ago, there appear to have been local calendars corresponding to local dialects. In classical Greece more than a hundred calendars have been identified, nor need we be astonished unduly by this multiplicity. Time was reckoned by the month as a basic unit, and the month, as the name implies, was determined according to the phases of the

[11] Cf. Philip, *The Calendar* (Cambridge: Cambridge University Press, 1921), p. 7.

[12] *Astronomy*, p. 184.

[13] Cf. Godfray, *A Treatise on Astronomy*, pp. 292-293.

[14] Cf. Philip, *The Calendar*, p. 9.—The Jewish calendar followed the same general principles of intercalation. Cf. *infra*, pp. 52-57, where more details are given with reference especially to the paschal cycle.

moon. It was with the naked eye that the moon was observed and a new moon—faint in the sky—was not always discerned on the same evening by people in different places. The same calendar might thus be variable in application. And in periods where sacred observances, as well as secular activities, were regulated strictly according to the calendar, the confusion—as Moslems have sometimes had reason to know — was regrettable. Some kind of uniformity in such calendars was thus desirable. In Palestine, after the exile, the Jews lit bonfires as beacons, beginning at Jerusalem, to signal the first day of the year and the first Sabbath of the month. The practice, we are told, only ended with the fall of Jerusalem, A.D. 70, and it illustrated the conditions under which even in a highly cultivated province of the Roman Empire the calendar had to be made available.[15]

Article 2—The Evolution of Our Own Calendar

The Gregorian calendar that is now used everywhere throughout the civilized world deserves special attention. That calendar, considered merely as a time-keeper, is of Egyptian origin. It was however handed down to us through Rome, a fact that accounts for the names of the months, their length, etc. A brief historical synopsis of the Latin calendar will not be out of place here.

§ 1—Before Caesar.

The calendar used by the Romans up to the time of Julius Caesar is a somewhat unknown quantity. It was of the lunisolar type and is said to have been the work of Numa Pompilius († 672 B. C.), the second king of Rome. There was no definite rule for the intercalations.

One thing, however, is morally certain. The Roman year, however long it might have been, began at one time on March 1st. That is clear to any one in the least familiar with Latin.

According to Macrobius (5th century A. D.) and Censorinus, the original Roman year contained 10 months and 304 days:

[15] *The Romance of the Calendar,* pp. 63-64.

6 months had 30 days (April, June, Sextilis,
September, November and December)—180 days
4 months had 31 days (March, May, Quintillis
and October)—124 days

304 days.

These same writers state that Numa added January and February at the end of the year. February was then the last month of the year and that accounts for the fact that it is shorter than the others and that the extra day in the leap years is added at the end of that month.[16]

In the course of time the calendar turned out as follows:

Martius	31 days	September	29 days
Aprilis	29 days	October	31 days
Maius	31 days	November	29 days
Junius	29 days	December	29 days
Quintilis	31 days	Januarius	29 days
Sextilis	29 days	Februarius	28 days.[17]

These 12 months accounted for 355 days. The balance was supplied by the intercalary month Mercedonius, which was wedged in after the 23rd day of February. This month originally consisted of 22 or 23 days in alternating sequence, but in the course of time it was given the last five days of February.

To make things worse the intercalation of Mercedonius was left to the wisdom of the pontiffs who dealt with the matter according to their professional secrets. Ignorance, negligence and, above all, corruption led to the most lamentable irregularities in the Roman calendar. When friendly magistrates were in office the years were invariably longer, and vice versa when the opposing faction held those positions. In Newcomb and Holden one reads:

[16] Cf. Philip, *The Calendar*, p. 9.—Other writers hold that January became the first month under Numa and that February was added at the end of the year. Cf. Ojetti, *Commentarium in Codicem Iuris Canonici, Normae Generales* (Romae: Apud aedes Universitatis Gregorianae, 1927), p. 196, note 2. This work will hereafter be referred to as *Normae Generales.*

[17] Cf. Wilson, *The Romance of the Calendar*, p. 107.

> It is said, for instance, that the Gauls having to pay a certain monthly tribute to the Romans, one of the governors ordered the year to be divided into 14 months, in order that the paydays might recur more rapidly.[18]

The only thing that was independent of the pontiffs in the Roman calendar was the enumeration of the days within the month. That followed definite and invariable rules. The Romans did not count the days as we do: January 1st, 2nd, 3rd, etc. They had what might be called the schoolboy attitude. A schoolboy does not count the days already gone by within the term. He looks forward to the next vacation and counts the days that separate him from that great event: so many days before Christmas, etc. The Romans had three definite dates that served as dividing lines within the month: the Kalends,[19] the Nones,[20] and the Ides.[21]

The Kalends were in every case the first day of the month. The other two varied a little. The four "menses pleni" (March, May, July, October), which had 31 days from the very beginning of the Roman calendar, had their Ides on the 15th. The other months had theirs on the 13th. The Nones being the ninth day before the Ides accordingly fell on the 5th or the 7th.

The day preceding any one of these fixed dates was called *Pridie,* as , for instance, *Pridiie Kalendas, Pridie Nonas, Pridie Idus.* The day that came before the *Pridie* was the third day before the Kalends, Nones or Ides.[22]

The Roman system of counting backwards, strange as it is, seems to have been well liked. The numerical system was to a

[18] *Astronomy,* p. 185.

[19] From the verb "calare." The pontiffs "called" the days in advance of the Kalends.

[20] From the Latin "nona," the ninth day before the Ides.

[21] The middle of the month: "Porro Idus diem qui medium dividit mensem: *iduare* enim Etrusca lingua dividere est."—Venerable Bede, *De Temporibus Liber,*—Migne, *Patrologiae Cursus Completus, Series Latina* (221 vols., Paris, 1844-1864), XC, 282. This collection will henceforth be referred to as *MPL.*

[22] Cf. page 14, where a complete Roman calendar is found.

The Roman Monthly Calendar

Days of the Month	March, May, July, October 31 days		January, August December 31 days		April, June, September, November 30 days		February 28-29 days		
1	Kalendis		Kalendis		Kalendis		Kalendis		
2	VI		IV	Ante Nonas	IV	Ante Nonas	IV	Ante Nonas	
3	V	Ante	III		III		III		
4	IV	Nonas	Pridie Nonas		Pridie Nonas		Pridie Nonas		
5	III		Nonis		Nonis		Nonis		
6	Pridie Nonas		VIII		VIII		VIII		
7	Nonis		VII		VII		VII		
8	VIII		VI	Ante	VI	Ante	VI	Ante	
9	VII		V	Idus	V	Idus	V	Idus	
10	VI	Ante	IV		IV		IV		
11	V	Idus	III		III		III		
12	IV		Pridie Idus		Pridie Idus		Pridie Idus		
13	III		Idibus		Idibus		Idibus		
14	Pridie Idus		XIX		XVIII		XVI		
15	Idibus		XVIII		XVII		XV		
16	XVII		XVII		XVI		XIV		
17	XVI		XVI		XV		XIII		
18	XV		XV		XIV		XII		
19	XIV		XIV		XIII		XI		
20	XIII		XIII		XII		X	Ante	
21	XII		XII	Ante	XI	Ante	IX	Kalendas	
22	XI		XI	Kalendas	X	Kalendas	VIII		
23	X	Ante	X		IX		VII		Leap Year
24	IX	Kalendas	IX		VIII		VI		VI
25	VIII		VIII		VII		V		VI
26	VII		VII		VI		IV		V
27	VI		VI		V		III		IV
28	V		V		IV		Pridie Kalendas		III
29	IV		IV		III				Pridie Kalendas
30	III		III		Pridie Kalendas				
31	Pridie Kalendas		Pridie Kalendas						

great extent based on the same principle. Thus 4 was not 3 + 1 but 5—1, IV, etc.

As one readily sees, the week had no place in the Roman calendar.[23]

§ 2—Julian Reform.[24]

The glaring imperfections of the Roman calendar were finally obviated by Julius Caesar, who became *Pontifex Maximus* in 63 B. C. In 47 B. C. when Caesar, with the help of the Alexandrian astronomer Sosigenes, set out to improve conditions, the calendar was out of joint to the extent that March had slipped into winter and June found itself in spring.

Two distinct problems faced Caesar. He had to set the calendar so as to make it agree with the sun, and then to regulate it in such a way as to make it keep good time in the future.

To solve the first problem he lengthened the year 46 B. C. by intercalating not only Mercedonius but two other months as well. That year therefore consisted of 15 months instead of 12, and of 445 days instead of 355:

Thus 12 normal months gave	355 days
Mercedonius gave	23 days
two extra months gave	67 days
	445 days.

Caesar's second step was to adopt a calendrical system that would eliminate the vagaries of intercalations. Sosigenes proposed the solar calendar then known in Egypt.[25] It was then believed that the tropical year was of 365 days and exactly 6 hours and not of 365 days, 5 hours, 48 minutes, and 46 seconds (or approximately so), as it really is.

The reformer therefore decreed that the year should be of 365 days with an intercalary day every fourth year. This leap year day was not added at the end of the year or even at the end of a month as should have been done. Superstition had a special

[23] Cf. *infra*, p. 35, where the week is considered *ex professo*.

[24] Cf. Wilson, *The Romance of the Calendar*, pp. 105-113.

[25] Cf. Philip, *The Calendar*, pp. 11-12.

weakness for Feb. 24, which had been the day set apart for intercalations for centuries. No new day, however, was added. Technically Feb. 24 was merely repeated. And Feb. 24 was, according to the Roman calendar, the sixth day before the Kalends of March (*sextus dies ante Kalendas*). The extra day became *Bis-sextus* or *bissextilis;* that term is still in current use to denote the leap year day in Latin, and also in French (année bissextile) and in Italian (anno bisestile).

Two years after the reform (44 B. C.) the name of the fifth month, *Quintilis,* was altered to July in honor of Julius Caesar.[26]

§ 3—Augustan Changes.[27]

Clear as Caesar's plans were, they were not properly carried out. The leap day was to be intercalated every fourth year, "*quarto quoque anno.*" The pontiffs unfortunately followed the Roman method of enumeration by which the number enumerated was inclusive both of the day from and the day to which the computation extended. Thus in practice the intercalation was made every third year.[28]

At any rate the intercalation of a leap day every third year continued for 36 years, during which 12 days instead of 9 had been added. The error was corrected by Augustus with the provision that the years from 9 B. C. to 3 A. D. should all be common years and that thenceforward the leap day should be added every fourth year.[29]

> Thus after the expiry of 48 years from the original introduction of the Julian Calendar the normal system was finally brought into operation. It may be noted, however, that chronologers have not recorded this error but have treated the leap years as having succeeded one another from the start.[30]

[26] Cf. Philip, *The Calendar*, p. 13.

[27] Cf. Wilson, *The Romance of the Calendar*, pp. 114-122.

[28] That method of reckoning also accounts for the fact that Our Lord's body is said to have rested three days in the tomb, even though He died on Friday afternoon and rose from the dead early Sunday morning.

[29] Cf. Philip, *The Calendar*, p. 15.

[30] Philip, *The Calendar*, p. 15.

A few other noteworthy changes have been attributed to the reign of Augustus. But the authenticity of some of these has been challenged in recent years. One thing, however, is certain. The name of one month was changed to Augustus to perpetuate the fame of the Emperor, just as Quintilis was changed to July to honor Julius Caesar who had been born in that month. The birthday of Augustus Caesar lay within September but he nevertheless chose Sextilis because that month had been most fortunate to his empire.[31]

Other months in the course of time were also given honorific imperial names. Thus April was for a while known as Neronius. Only July and August, however, have been handed down to us.

Another point worthy of mention is the fact that Sextilis had 30 days in the old Roman calendar, and that it certainly had 31 days after the Augustan changes. It is commonly believed that the Emperor subtracted a day from February and added it to his own month. This led to a new arrangement of the calendar. To summarize, the Julian calendar was modified by Augustus as follows:

Month	Julian	Augustan	Month	Julian	Augustan
January	31	31	July	31	31
February	29-30	28-29	August	30	31
March	31	31	September	31	30
April	30	30	October	30	31
May	31	31	November	31	30
June	30	30	December	30	31.[32]

These changes were evidently unwise and only served to emphasize the irregularities of the calendar. Yet this arrangement has endured to our own day.[33]

§ 4—Gregorian Reform.[34]

[31] Cf. Wilson, *The Romance of the Calendar,* p. 119.

[32] Cf. Wilson, *The Romance of the Calendar,* p. 121.

[33] For more details concerning the Augustan changes cf. Philip, *The Calendar,* p. 14; Lamont, "Augustan Myths,"—*The Journal of Calendar Reform,* IX (1939), 10-13.

[34] Cf. Wilson, *The Romance of the Calendar,* pp. 137-144.

The Julian calendar was a nearly perfect time-table and was of the greatest utility for the administration of the far flung Roman provinces. It enjoyed a distinct advantage over our own calendar inasmuch as all the feasts and celebrations always fell on the same day of the month. The week did not exist as a calendrical entity and there were no Sundays and no movable feasts, i. e., feasts without any fixed correspondence to the dates of the calendar.

However, the Julian calendar was based on a slight astronomical error. March 24 had been designated as the date of the vernal equinox, but as time flowed on the vernal equinox was gradually slipping into the winter months. Sosigenes in fact had considered the tropical year as of 365 days and exactly 6 hours. In reality the true year lasts 365 days, 5 hours, 48 minutes and 46 seconds. That meant that Caesar's year was 11 minutes and 14 seconds too long. In a century the error mounts up to 18 hours, 43 minutes and 20 seconds; and after a thousand years we have a total of 7 days, 19 hours, 13 minutes, 20 seconds. When the error was finally eradicated by Pope Gregory XIII in 1582 the calendar had gained close to 13 days (45 B. C. to 1582 A. D.—1627 years multiplied by 11 minutes and 14 seconds). But, as shall be explained in a moment, Gregory XIII did not make the correction as retroactive to 45 B. C., but to 325 A. D., the year of the Ecumenical Council of Nicaea. That accounts for the fact that 10 days only, and not 13, were dropped out of the year 1582.

Calendar reform was seriously studied and examined in the 13th century when Joannes Campanus and Roger Bacon made suggestions to the Holy See. From then on changes were periodically advocated until the Council of Trent. This august assembly was too busy with more urgent questions, and in the last session left the matter to the Vatican. Gregory XIII was the first pope able to devote the required time to the reform.

Various plans were submitted, but the one finally accepted was proposed by Antonio Giglio (Lilius).[35]

[35] The plan was previously drawn up by Antonio's brother, Luigi († 1576).

The approval of the plan by an international commission was followed by letters sent to the Catholic Princes (Jan. 5, 1578) for their support. Finally the introduction of the reform was considered for 1581 in spite of the opposition of the University of Paris. But various difficulties arose and the Bull promulgating the new calendar was not issued before Feb. 24, 1582.[36]

That document, known as the Bull "Inter gravissimas," contained two principal provisions:[37]

a—It restored the date of the vernal equinox to March 21, its presumed date at the time of the Council of Nicaea in 325 A. D.[38] In 1582 the actual vernal equinox fell on March 11. The civil calendar was therefore 10 days ahead of the celestial timekeeper. It was therefore decided that 10 days should be dropped from the calendar so as to bring March 11 (vernal equinox) up to March 21; the decree stipulated that Oct. 4 would be followed by Oct. 15.[39]

b—In order to regulate the calendar so as to make it agree with the celestial clock it was provided that 3 out of 4 centurial years should be common years instead of leap years as under the Julian

[36] Cf. Pastor, *The History of the Popes from the Close of the Middle Ages* (29 vols., translation, vols. I-VI ed. by Frederick I. Antrobus; Vols. VII-XXIV, ed. by Ralph Kerr; Vols. XXV-XXIX, ed. by Dom Ernest Graf, St. Louis: Herder, 1906-1938), XIX, 283 ff.

[37] *Bullarium Diplomatum et Privilegiorum Sanctorum Romanorum Pontificum, Taurinensis Editio* (24 vols. et Appendix, Augustae Taurinorum-Neapoli, 1857-1872), VIII, 386.

[38] The reason why Gregory XIII did not adopt March 24 of the Julian calendar was, in all likelihood, because March 21 had become the universal date according to which Easter was reckoned. Concerning the rôle of the Council of Nicaea with reference to the determination of Easter cf. Hefele-Leclercq, *Histoire des Conciles d'après les Documents Originaux* (10 tomes in 19 vols., Paris, 1907-1938), Tome I, Part I, 450-477; Schwegler, "The Vatican and Calendar Reform,"—*The Homiletic and Pastoral Review,* XXXIV (1934), 804. This publication will henceforward be referred to as *HPR.*

[39] A name often mentioned in connection with the reform of the calendar is that of St. Teresa of Avila. This great saint died Oct. 4, 1582, the last day of the Julian calendar. Cf. Cicognani, *Canon Law,* p. 670, note 18.

calendar. The only centurial years retained as leap years were the ones divisible by 400 according to the following table:

1600 leap year	1700	1800	1900 common years
2000 leap year	2100	2200	2300 common years
2400 leap year	2500	2600	2700 common years.

According to that plan the average Gregorian year contains 365.2425 mean solar days or 365 days, 5 hours, 49 minutes and 12 seconds, which is about 26 seconds longer than the true solar year:

Gregorian year	365 d.	5 h.	49 m.	12 s.
Solar year	365	5	49	46.

In other words the Gregorian calendar gains a day every 3323 years. Sir John Herschel suggested a further correction according to which the year 4000 and its multiples would be considered common years.[40]

The Gregorian reform was not accepted by all the Western World at the same time. This divergence only aggravated the already none too simple problem of time-keeping.

As could be imagined, the Papal States immediately adopted the new calendar. So did Philip II (1556-1598) of Spain. France came in line in December of the same year. The other catholic countries and the catholic sections of Switzerland, of the Netherlands and of the German Empire did not follow suit until 1583.

[40] Cf. Philip, *The Calendar*, p. 23. "The Persian method of interpolation deserves to be mentioned. It was introduced in the 11th century, that is, five centuries before the Gregorian reformation, and is even more exact:—Three years of 365 days are followed by a year of 366 days, as in the Julian calendar; this is done seven times in succession, but the eighth period consists of five years, four common years followed by a year of 366 days. Thus in 33 years there are 8 leap years. Therefore, the average length of the Persian civil year is 365 8/33 days, or 365.242424 days, which is in excess of the present length of the tropical year by .000208 days, or 1 day in about 5000 years."—Godfray, *A Treatise on Astronomy*, p. 292, note. Cf. *infra*, p. 21, where the Gregorian reform is graphically explained.

"Vernal equinox in the the time of Numa, about 700 B. C. .... March 24, 46 B. C.

It is here seen by the errors of the Julian Calendar, the Vernal Equinox is maďe to occur three days earlier every 400 years, so that in 1582 it fell on the 11th instead of the 21st of March."

Day of March	Julian Calendar	Gregorian	Year
23			1 A. D.
22			100 " "
21			300 " "
20			400 " "
19			500 " "
18			600 " "
17			800 " "
16			900 " "
15			1100 " "
14			1200 " "
13			1300 " "
12			1400 " "
11	By suppressing 10 days, Coincidence	Restored in	1600 " "
	Hours behind time, 18	6 in advance	1700 " "
	" " " 12	12 " "	1800 " "
	" " " 6	18 " "	1900 " "
	Coincidence	Restored	2000 " "
	Hours behind time, 18	6 in advance	2100 " "
	" " " 12	12 " "	2200 " "
	" " " 6	18 " "	2300 " "
	Coincidence	Restored	2400 " "

Vernal Equinox at the Council of Nice 325 A. D.

Restored by Julius Caesar 46 B. C., to the place it occupied in the time of Numa.

"By the Gregorian rule of intercalation the coincidence of the solar and civil year is restored very nearly every 400 years."[41]

[41] Packer, *Our Calendar* (Corning, N. Y., 1893), pp. 59-60.

The Protestant States of Germany abandoned the Julian calendar in 1700.[42]

England, and consequently the United States, did not adopt the Gregorian calendar before 1752. September 2, 1752 was followed by September 14.[43]

This change must therefore be borne in mind when we deal with English or American dates prior to 1752. Thus, George Washington was born Feb. 11, 1732, according to the Julian calendar (or Old Style, as it is sometimes called). However, we celebrate his birthday on Feb. 22, which is the Gregorian, or New Style, date corresponding to Feb. 11 of the Julian calendar.

Where accuracy is desired, ambiguous dates are written in a special way. Thus Queen Elizabeth died March 24, 1602, Old Style, or April 3, 1603, New Style.[44] This date would sometimes be given thus:

24 March 2
.... 160-
3 April 3.[45]

Sweden adopted the Gregorian calendar in 1753. The following are the more recent changes: Japan, 1873; China, 1912; Bulgaria, 1916; Soviet Russia, 1918; Yugoslavia, 1919; Roumania and Greece, 1923; Turkey, 1927.[46]

[42] Cf. Pastor, *The History of the Popes,* XIX, 283 ff. These dates for the adoption of the Gregorian calendar in the various Catholic countries and sections here mentioned are not given alike by all authors. Cf. Cappelli, *Cronologia*, pp. 29-31.

[43] The Julian calendar observed a leap year in 1700, but not the Gregorian calendar. This extra day added to the 10 already existing in 1582 made a difference of 11 days between the two calendars. That accounts for the suppression of 11 days here as opposed to 10 in 1582.

[44] The difference of a whole year (1602-1603), which is here noticed, is due to the fact that the first day of the year, according to the Old Style reckoning, was March 25 and not January 1. Cf. *infra,* p. 24.

[45] Cf. Gerard, "Chronology,"—*The Catholic Encyclopedia* (15 vols., New York, 1913-1914), III, 739. This work will henceforth be referred to as *CE.*

[46] Cf. *The Journal of Calendar Reform,* IX (1939), 65. Here too Cappelli offers some dates different from the ones presented. Cf. *Cronologia,* pp. 30-31.

This, however, does not mean that these countries adopted the Gregorian calendar thoroughly. In many cases the use of the new calendar is limited to international affairs, while the Julian calendar still holds for internal computations.[47]

If one now considers the use of the Gregorian calendar in church affairs, one finds that it has been adopted by the great majority of the Oriental rites besides the Latin.

> The Julian reckoning is, however, still followed by the Catholic Bulgarians and Ruthenians, and by the Orthodox patriarchates of Antioch, Alexandria, Jerusalem and Serbia, the Nestorian, Gregorian, Armenian, Jacobite and other churches (Roumania, Greece and Constantinople have accepted the Gregorian reckoning for fixed feasts). These are consequently thirteen days behind and the celebration of Easter and other feasts differs accordingly.[48]

Article 3—Accidental Details

Thus far in the explanation of the calendar the writer has adhered to the substance of the thing and has ignored all extrinsic considerations. A few of these accidental details, however, are worthy of mention.

§ 1—The Beginning of the Year

It is certain that January 1 was not always the first day of the year. The logical day on which to begin the civil year would be either one of the equinoxes or one of the solstices. These are the days marked out by nature itself. They mark the various turning points of the earth's revolution around the sun. But the makers of our own calendar have chosen other days, if one excepts perhaps Numa Pompilius, who gave Rome its first orderly calendar. March 1 was the beginning of the Roman civil year and most likely corresponded with the vernal equinox.

Many opinions are put forth as to the time that January 1 became the first day of the year. Some writers mention the year

[47] Cf. Wilson, *The Romance of the Calendar,* p. 326; Turano, "The Calendar is out of Date,"—*The Reader's Digest,* XXXIV (1939), No. 203, pp. 95-96.

[48] *The Catholic Encyclopoedic Dictionary* (London: Cassell, 1931), s. v. *Calendar, The Julian.*

153 B. C.[49] Others refer to the year 45 [46] B. C., when Julius Caesar reformed the calendar.[50]

Throughout the Middle Ages January 1 was not used as the starting point of the year by Christendom.[51] Various religious feasts have at one time or another been used by the same country. The two feasts more often used were March 25 (Style of the Incarnation) and December 25 (Style of the Nativity). The following will give a rough idea of the dates actually adopted:

England used March 25 or December 25 before the Norman Conquest (1066),
January 1 from 1087 to 1155
March 25 from 1155 to 1752.

France used Christmas Day, Easter Eve or March 25 until 1564 when January 1 was adopted.

Germany anciently used Christmas. January 1 was adopted in 1544.

Rome and the greater part of Italy used December 25 until 1582

[49] Cf. Wilson, *The Romance of the Calendar,* p. 107; *The Encyclopaedia Britannica* (14. ed., 24 vols., The Encyclopaedia Britannica Company, Ltd.: London, 1936), s. v. *Calendar.*

[50] Thus Gerard, "Chronology,"—*CE,* III, 739. In his information column "Answers to Questions" (The Evening Star [Washington, D. C.], January 27, 1940) Frederic J. Haskin asserts that January 1 was accepted as the beginning of the year as early as 251 B. C. According to Wilson (*The Romance of the Calendar,* p. 296), Julius Caesar, when he reformed the calendar, wanted to start the year with the winter solstice, but was prevented from so doing by popular opinion. In deference to this opinion he chose the new moon which was nearest to the solstice. That year (46 B. C.) it fell on January 1. According to another theory Numa added January and February at the beginning and not at the end of the year.

[51] That is, for chronological purposes. In ordinary affairs January 1 was still New Year's Day. Cf. Gerard, *loc. cit.*—"Neque tamen obstitit, quin Kalendae Januariae primus anni dies semper habitae sunt. Extat enim Charta Drogonis Ambianensis Dom. *de Vinacourt,* in Tabulario Vicedomini Ambian, fol. 69. cum hoc temporis adscripto charactere: *Fait en l'an de l'Incarnation de Notre Seigneur Jésus-Christ* 1183 *el mois de Janvier lendemain du premier jour de l'an.*"—Du Cange, *Glossarium ad Scriptores Mediae et Infimae Latinitatis* (2. ed., 6 vols., Paris, 1733-1736), s. v. *Annus* (I, 462).

when January 1 was introduced.[52] Scotland adopted January 1 in 1600.[53] "In Venice," writes Wilson, "the day was March 1. In Denmark the year, for a time, began with the feast of Saint Tiburce—August 11th." [54] "From 1500 to 1725, Russia started her year with the spring equinox, March 21. The French revolutionists began their year with the autumnal equinox, September 21."[55]

In Alexander Philip's *The Calendar* one reads:

> With the Egyptians the commencement (of the year) was made at the autumnal equinox—the reason probably being that that date coincided with the greatest height of the Nile Flood—(Herodotus II, 19,)—to them the most outstanding natural event in the year. Very probably from them the Jews derived the custom of dating their year also from the autumnal equinox. To this day the Jewish civil year commences with the month of Tisri [corresponds approximately to September]. But ever since the deliverance from Egypt the ecclesiastical year of the Jews has commenced at the vernal equinox with the month Nisan [corresponds to March].[56]

§ 2—The Meaning of "1941"

Another interesting angle in the study of chronology is the number of the year in which we are. Thus our calendar bears the number 1941. What event marked the beginning of this cycle? Why and since when are dates reckoned from that event? Have other events given birth to other eras? Such is the nature of the problem that now confronts us.

[52] Cf. Gerard, "Chronology," *CE,* III, 739.

[53] Cf. *The Encyclopaedia Britannica,* s. v. *Calendar.*

[54] *The Romance of the Calendar,* p. 297.

[55] *Op. cit.,* p. 293.

[56] Page 8.—Cf. also Cappelli, *Cronologia,* pp. 8-22, where one finds a list of the particular style adopted by the various States in Europe.—The formulas more frequently used were the following: *anno incarnationis, ab incarnatione Domini, Dominicae incarnationis, trabeationis, ab incarnati verbi misterio, anni Domini, a nativitate Domini, orbis redempti, salutis gratiae, a passione Domini, a resurrectione, Circumcisionis, etc.*—Cf. *op. cit.,* p. 9—

Theoretically the year 1472 *ab incarnatione Domini* should have run from March 25, 1472, to March 25, 1473, of our reckoning. In practice, however, a year of the Incarnation might have begun December 25. Cf. *op. cit.,* p. 9, note 7.

A year is not a separate independent unit. It is part of the continuous flow of time, it is but a section of the chain that links the history of man. Today it is immaterial to the ordinary individual whether Leo I (†461) dated a document under the consulate of Maximus and Paternus or any other consuls. And had we no other way of reckoning, confusion might go so far as to imagine Caesar a contemporary of Alexander the Great († 323 B. C.). A buoy is useless unless it be firmly attached to the bottom of the river or channel. So must a calendar be anchored if it is to mean something. The calendrical anchor is attached to some important event that is firmly imbedded in the memory of men. Maximus and Paternus may have been famous men in their time, but today they have been forgotten and their chronological whereabouts is unknown unless we be told that they were consuls in the year 443 A. D.

Very few events have proved capable of drawing the attention of all men for a considerable length of time. Among these, two are of singular importance: the creation and the nativity of Christ.

The creation of the world, owing to our lack of knowledge concerning its exact date, has long been abandoned as a calendrical entity. On the other hand the influence of the Saviour's birth in the field of chronology is still gaining momentum. The so called "Christian Era" has never been more widely used than it is today.

Gerard lists the following epochs among the better known besides the Christian era:[57]

1—Era of Abraham, from October 1, 2016 B. C.
2—Era of the Olympiads, July 13, 776 B. C.[58]
3—Era of the Foundation of Rome, April 21, 753 B.C.[59]

[57] "Chronology," *CE,* III, 740. Numerous have been the chronological systems reckoning the years from the beginning of the world. Among these the best known is the Byzantine, or Greek era, which originated in the 7th century A. D., no one knows exactly where. According to this system the creation dated back to the year 5508 B.C. The year 1 A.D. corresponds in part to the year 5509 of this era. September 1 was the beginning of these Byzantine years. In Russia this system was abolished January 1, 1700, by Peter the Great. Cf. Cappelli, *Cronologia,* p. 3.

4—Era of Nabonassar, February 26, 747 B. C.[60]
5—Era of Alexander, November 12, 324 B. C.
6—Greek era of Seleucus, September 12, 312 B. C.[61]
7—Era of Tyre, October 19, 125 B. C.
8—Caesarian era of Antioch, August 9, 48 B. C.[62]
9—Julian era, January 1, 45 [46] B. C.
10—Era of Spain or of the Caesars, January 1, 38 B. C.[63]
11—Era of Augustus, September 12, 31 B. C.[64]
12—Egyptian year, August 29, 26 B. C.[65]
13—Era of Martyrs or of Diocletian, August 29, 284 A. D.[66]
14—Era of the Armenians, July 9, 552 A. D.[67]
15—Mohammedan era, July 16, 622 A. D.[68]
16—French Revolutionary eras—

A letter sent by Iwan III Wassiliewitch to the duke of Milan, Gian Galeazzo Maria Sforza, is thus dated: "Datae Moscoviae, mense aprilis annis a constitutione mundi uno et septemmilibus." Cf. *op cit.*, p. 3, note 1. A similar system is also found in the footnotes of the Douay version of the Bible.

58 An Olympiad was a cycle of four years and was used by historians, but very seldom found in documents of state or inscriptions. The first six months of 1 A. D. correspond to the last six months of Olympiad CVC-1. The last six months of 1 A. D. correspond to the first six months of Olympiad CVC-2. The Olympiads continued to 396 A. D. (Olympiad CCXCIII).

59 This era was probably not adopted before the second century B. C.

60 Also called the Babylonian era. It is the basis of the calculations of Ptolemy.

61 Also called the Macedonian era. It is the starting point of the chronological system used in the Books of the Machabees and known as the era of the Kings.

62 Instituted to commemorate the Battle of Pharsalia.

63 Dates from the conquest of Spain by Augustus. This era was extensively used in the Iberian peninsula and in southern France throughout the Middle Ages, and was employed long after the rest of Christendom had adopted the Christian era.

64 Instituted to commemorate the Battle of Actium.

65 Instituted on the reformation of the Egyptian calendar by Augustus.

66 Employed by Eusebius and early ecclesiastical writers. This era was used especially in Egypt, and is actually followed by the Copts (upper Egypt). Cf. Cappelli, *Cronologia,* p. 5.

Era of Liberty, January 1, 1789 A. D.
Republican eras, January 1, 1792 A. D.
September 22, 1792 A. D.

According to Philip: "In India an era known as the Kaliyug has been employed, its assumed commencement being the year 3102 B. C. In China the commencing date seems to have been 2397 B. C." [69]

Other shorter periods have also found favor throughout the ages, namely, the regnal years and the indictions.

The regnal years indicated the year of the reigning prince. That practice was undoubtedly handed down from the Roman custom of giving the name of the consuls in office. A vestige of this is yet to be found in papal documents. Thus one reads at the end of Pope Pius' Encyclical Letter to the Church in the United States, *"Sertum Laetitiae"*: "Given at the Vatican, on the Feast of All Saints, in the Year of Our Lord 1939, the first of Our Pontificate." [70] The Fascist era is fairly well known at the present day. An official date would thus be expressed in Italy:
January 30, 1940—E. F. XVIII.

The indiction was a cycle of 15 years introduced by Constantine in 312 A. D., and was originally intended to be used in connection with the assessment of taxes and the public accounts. These cycles began January 1, 313.[71]

There now remains the question of the Christian era. The words "Anno Domini" are familiar to us, but they were not known to a great extent a thousand years ago. The precise date of the birth of Christ and the year when the Christian era was introduced are still shrouded in mystery. One can hardly state anything beyond the fact that this era came into general use only towards

[67] Commemorates the consummation of the Armenian schism by their condemnation of the Council of Chalcedon.

[68] Dates from the Hegira or Flight of Mohammed from Mecca and his entrance into Medina.

[69] *The Calendar*, p. 49.

[70] Official Vatican English Translation as found in the National Catholic Welfare Conference publication.

[71] Cf. Philip, *The Calendar*, p. 46; Wilson, *The Romance of the Calendar*, p. 134.

the tenth century, even though it had been proposed in the sixth century (532) by Dionysius Exiguus, to whom is attributed the calculation of the year of Christ's birth.[72] The works of the Venerable Bede († 735) were influential in the adoption of the Christian era.[73]

It is recognized today that Dionysius erred in his computations. He inferred that Christ was born December 25, 753 A. U. C., and since the Roman civil year began January 1 (7 days later) it was found more convenient to make the Christian year begin on the same day. The year 1 A. D. thus coincided with the year 754 A. U. C., and 1 B. C. corresponded to 753 A. U. C. Dionysius came to this conclusion by accepting the widespread tradition that Christ was born in the 28th year of the reign of Augustus.[74]

Alexander Philip thus explains the computation:

> Dionysius, however, fell into error in computing the commencement of that reign which he assumed to be 727 A. U. C., the year in which Octavius adopted the name or title of Augustus, whereas in point of fact his reign was always computed from the date of the Battle of Actium, 2nd September, 732 A. U. C. (31 B. C.). The position can be best understood by reference to a tabular statement of corresponding dates commencing with the Battle of Actium which we here subjoin.

2nd Sept. 723 A.U.C.	=date of the Battle of Actium.
Sept. 723-Sept. 724	=1st year of Augustus.
∴ Sept. 750-Sept. 751	=28th year of Augustus.
∴ 25th Dec. 750	=Birth of Christ=4 B.C.
1st Jan. 751-31st Dec. 751	=1 Anno Christi=3 B.C.
1st Jan. 754-31st Dec. 754	=4 Anno Christi=1 A.D.[75]

[72] The first application of this era in documents occurred in England in 704. The Apostolic Chancery did not adopt this system before the pontificate of John XIII (965-972). From then on its use became widespread in Europe. Cf. Cappelli, *Cronologia,* p. 9.

[73] Cf. Wilson, *The Romance of the Calendar,* p. 135.

[74] That is the statement of Clement of Alexandria (*Stromata,* I, 21) according to Philip, *The Calendar,* p. 51. Cf. Collins, "Can the Star of the Magi give us the Date of Christ's Birth,"—*The Ecclesiastical Review,* CI (1939), 551-555. This publication will hereafter be referred to as *ER*.

[75] *The Calendar,* p. 51.

One is to note, in passing, the makeshift terminology of our dating system. The abbreviation B. C. stands for the English words Before Christ, while the letters A. D. stand for the Latin *Anno Domini.*

§ 3—Papal Documents

The preceding pages have given a rough idea of the state of confusion that existed at one time in matters chronological. Our present day system has been in prevalent use for well over two centuries. The use of the Christian era gained considerable popularity from the tenth century on. The Gregorian reform of 1582 not only adjusted the calendar, but also marked the beginning of a new epoch in the adoption of January 1 as the first day of the year.

In this last respect, however, vestiges of the old style have endured to the present century. Many papal documents, the Bulls especially, have been dated according to the style of the Incarnation for many years after the use of January 1 as the first day of the year had become practically exclusive in other affairs. Thus the Constitution *Officiorum et Munerum* is actually dated VIII kal. febr. 1896, which corresponds to January 25, 1897.[76]

In 1908 Pius X prescribed that thenceforth all papal documents be dated from January 1.[77] The old system was nevertheless actually used until 1913.[78]

Article 4—The Component Parts of the Calendar

It is now proper to consider the various time-units that the calendar metes out to us, namely the year, the month and the week.

[76] Cf. Vermeersch-Creusen, *Epitome Iuris Canonici,* Vol. I, (5. ed., Mechliniae: Dessain, 1933), n. 84, 1. Hereafter this work will be cited as *Epitome.*

[77] "In posterum vero in omnibus Apostolicis litteris, sive a Cancellaria sive a Dataria expediendis, initium anni ducetur, non a die Incarnationis Dominicae, hoc est a die XXV mensis martii, sed a kalendis ianuariis."—Constitutio *Sapienti Consilio,* 29 iun. 1908, § III, n. 5—*Codicis Iuris Canonici Fontes* (9 vols., Romae [later Civitate Vaticana]: Typis Poly-

§ 1—The Year

The fundamental and outstanding unit among the three is the year, for it is the only one according to our actual solar calendar that is regulated by the motion of the celestial bodies. Its length is not merely arbitrary and conventional as is the case for the month and the week.

Now, the word year has in itself a rather loose meaning. It simply refers to a cycle or ring of some kind. More specifically, it denotes the motion of a body around another and back again to the starting point. It corresponds to the Latin *annus,* which comes from *annulus,* ring.[79] Thus the divers planets have years of varying length according to the time they take to accomplish one complete revolution around the sun.

Even if one reserved the term "year" to the earth's peregrinations, one would be confronted with years of divers kinds and lengths. "A year," writes Godfray, "is the period of the earth's revolution around the sun, from some determinate position back again to the same." [80] Change the starting point and you will have a different kind of year. Besides the civil year which has already been mentioned, there are three kinds of year worthy of note: the sidereal, the anomalistic, the tropical.

"The sidereal year, as its name implies, is the time occupied by the sun in apparently completing the circuit from a given star

glottis Vaticanis, 1923-1939), n. 682. Reference to this work will hereafter be made by the word *Fontes.*

Much useful information concerning the manner of dating in papal documents is found in Jaffé, *Regesta Pontificum Romanorum ab Condita Ecclesia ad Annum post Christum Natum* MCXCVIII (2 ed., 2 vols., Lipsiae, 1881-1888), II, VIII-IX. Cf. also Covarruvias, *Opera Omnia* (2 vols., Coloniae Allobrogum, 1679), Variarum Resolutionum, lib. I, cap. XII, n. 1; Ojetti, *Normae Generales,* p. 195, note 10.

[78] Cf. Vermeersch-Creusen, *Epitome,* I, n. 84, 1.

[79] Cf. Antonellus, *Tractatus Novissimus et Absolutissimus de Tempore Legali* (Venetiis, 1692), lib. I, cap. III, n. 11. This work will hereafter be cited by the words *De Tempore Legali.* The name of this author is also given as Antonelli. Cf. also Virgil, *Georgics* (*Bucoliques-Georgiques, Enéide,* ed. J. B. Lechatellier [12. ed., Paris: De Gigord, 1924]), II, 402: "Atque in se sua per vestigia volvitur annus."

[80] *A Treatise on Astronomy,* p. 158.

to the same star again."[81] According to Young this is the true year from the mechanical point of view.[82] It corresponds to one complete revolution of our sphere (360°) about the sun. That revolution takes 365.256374 mean solar days, or 365 days, 6 hours, 9 minutes, and 9 seconds.

In the anomalistic year the starting point is the perihelion of the earth's orbit. The perihelion, incidentally, is the point where the earth is nearest to the sun, and is opposed to the aphelion, the point where the earth is remotest from the sun. The cycle from the perihelion to the perihelion takes approximately a solar year. It consists of 365.259641 days, that is, 365 days, 6 hours, 13 minutes, and 48 seconds. It corresponds to 360° 11′ 25″.[83] It is nearly half an hour longer (25 minutes and 2 seconds) than the tropical year, so that the perihelion, which now occurs on January 2, is always slipping back in the calendar:

Thus, in year 4000 B.C. the perihelion occurred on September 21
" " 1250 A.D. " " " " December 21
" " 6590 A.D. " " will occur on March 21
" " 11910 A.D. " " " " " June 22
" " 17000 A.D. " " " " " September 21.

A complete cycle is made in approximately 210 centuries.[84]

The tropical year is the one our calendar is based on. This year determines the beginning of the seasons and all the important phenomena of vegetation and life. It is the unit marked out by nature for the use of man. "The tropical year," writes Young, "is the time included between two successive passages of the vernal equinox by the sun.[85] Thus it may be seen that January 1 is nothing but a conventional date for the beginning of the civil year. The natural and logical beginning would be either one of the equinoxes or one of the solstices.

The tropical year, also called the equinoctial year, does not, as is sometimes believed, mark the exact interval of time taken by the earth to complete a revolution around the sun. There is a

[81] *Loc. cit.*
[82] *Elements of Astronomy* (revised ed., Boston, 1903), p. 84.
[83] Cf. Godfray, *A Treatise on Astronomy,* p. 158.
[84] Cf. Flammarion, *Astronomie Populaire,* p. 15.
[85] *Elements of Astronomy,* p. 84.

little difference. When the tropical year is finished, the earth must still revolve for 20 minutes and 23 seconds before it completes a full revolution (360°) around the sun. The tropical year consists of 365.242216 mean solar days, or 365 days, 5 hours, 48 minutes and 46 seconds. This corresponds to 360° less 50′ 22″.

In the way of recapitulation the following figures may be given:

Sidereal year	365.256374 d. or 365 d 6 h 9 m 9 s = 360°
Anomalistic year	365.259641 d. or 365 d 6 h 13 m 48s = 360° 11′ 25″
Tropical year	365.242216 d. or 365 d 5 h 48m 46s = 360° less 50′ 22″.[86]

§ 2—The Month

The month, as it is conceived today, is merely an arbitrary division of the year and corresponds to either 28, 29, 30 or 31 days, as has been shown in the consideration of the evolution of the Gregorian calendar. Formerly the month was measured by the motion of the celestial bodies and was the interval of time between two successive new moons, just as the year is today the interval of time between two successive vernal equinoxes.

Month and moon are closely associated. Month is a general term and means some sort of lunar cycle, just as the word year refers to a cycle of the earth. There are different kinds of months according to the angle from which one examines the moon's peregrinations.

The moon revolves around the earth in approximately 27 days and 8 hours. These figures vary according to the method of calculation. One reads in Wilson:

> There is the sidereal month (27.32166 days) based on the relation between the earth, the moon and the far distant stars. It is virtually the same as the tropical month (27.32158 days) based on the equinox. The anomalistic month is slightly longer (27.55455 days). It is the period that elapses between what is called perigee and perigee—which simply means the moments when the moon, in its slightly elliptical orbit, is nearest the earth. Another month is known as nodal or diaconic (27.21222

[86] Cf. Newcomb and Holden, *Astronomy*, p. 154; Godfray, *A Treatise on Astronomy*, p. 158.

> days), and this month is more delicately adjusted. The orbit of the moon is in one plane. The orbit of the earth is in another plane. Where these planes intersect there is a straight line pointing to the nodes, and the nodal month is reckoned relatively to this line.[87]

Now, all these months are of importance to astronomers, but not to calendar builders. There is another month, called synodical, which is visible to everyone and regulated by the various phases of the moon. The earth and the moon behave in such a way throughout the year that the moon is sometimes between the earth and the sun. At that particular point the moon is lost in the sun's rays and cannot be seen. That is the new moon. That celestial body gradually moves away from the sun and becomes more and more visible until it reaches the opposite side of the sun. We then have the full moon. From then on visibility decreases to zero, when we again have a new moon.

The interval of time between two successive new moons is 29.53059 mean solar days, or 29 days, 12 hours, 44 minutes, and 29 seconds. The various phenomena observed during this synodical month are commonly known as the moon's phases. Thus we have the new moon, the first quarter, the full moon, the last quarter, and again the new moon, and so on, indefinitely.

The difference in length between the synodical month and the others might at first be intriguing. If the earth were not revolving around the sun, the sidereal and synodical months would have exactly the same value; 27 1/3 days. In reality, during one lunation the sun moves east among the stars to the extent of approximately 30°. The moon has therefore to travel 360° + 30°, or 390°, to overtake the sun: 360° are completed in 27 1/3 days, and 390°, in 29½ days.

On this subject it might be opportune to say a word or two about the lunar day. This time-unit is the interval of time between two successive transits of the moon over the same meridian, and is reckoned similarly as the solar day.

During one lunation the sun crosses the meridian 29½ times and the moon 28½ times. "The moon," says Godfray, "crosses

[87] *The Romance of the Calendar*, pp. 49-50.

the meridian about 50½ minutes later every day."[88] And that explains the fact that the moon does not always set and rise at the same place in the heavens as does the sun. "These variations depend on the latitude of the place: at Cambridge [England] the retardation may amount to 1 hour and 15 minutes, and at other times be only 18 or 20 minutes."[89]

The phases of the moon are easily observed and therefore constituted with the day the earliest time-units.

§ 3—The Week

The week is the most intriguing time-unit for study and observation. It holds a prominent place in our calendar, and plays a still more important part in our every-day routine. The first or the second day of the month as such mean very little to us, but the first day of the week means to the great majority of civilized men a day of rest, a day consecrated to God. The week as it is known today is undoubtedly of Judeo-Christian origin, but it is nearly impossible to say precisely when and where it originated as a component part of the calendar.

> In the Roman chronoligical system of the Augustan age the week as a division of time was practically unknown. . . . In the course of the first and second century after Christ, the hebdomadal or seven day period became universally familiar, though not immediately through Jewish or Christian influence.[90]

Market weeks of divers lengths have been used throughout the world since the dawn of time. The following may be given in the way of illustration:

3 days among the Muysca of Bogota—
4 days among the West African tribes—
5 days in Central America, the East Indian Archipelago and old Assyria—
6 days among a tribe in Togo—
8 days among the ancient Romans—[91]

[88] *A Treatise on Astronomy*, p. 242.
[89] *Loc. cit.*
[90] Thurston, "Calendar, Christian,"—*CE*, III, 158.
[91] Cf. [Nilsson], "Calendar,"—*The Encyclopaedia Britannica*, IV, 575.

The Roman week was called the *"nundinae"* and is said to have been instituted by Servius Tullius († 534 B. C.), the 6th king of Rome. These market days had a religious significance, and work was often prohibited on their account, especially in Africa.[92]

A seven day week was known in Egypt and might have been suggested by the phases of the moon, or again by the number of planets known in ancient times, an origin which is rendered probable from the names universally given to the different days of which the week is composed.[93]

Here are the Latin names of the days of the week with the corresponding English and Saxon words:

Latin	*English*	*Saxon*
Dies Solis	Sunday	Sun's day
Dies Lunae	Monday	Moon's day
Dies Martis	Tuesday	Tiw's day
Dies Mercurii	Wednesday	Woden's day
Dies Jovis	Thursday	Thor's day
Dies Veneris	Friday	Frigg's day
Dies Saturni	Saturday	Saterne's day.[94]

[92] Cf. Philip, *The Calendar,* p. 19; Toso, *Ad Codicem Iuris Canonici Commentaria Minora,* Vol. I, (Tiferni Tiberini: Typographia Vinciana, 1921), p. 97. This work will hereafter be referred to as *Commentaria Minora.*

[93] "In the Egyptian astronomy," writes Packer, "the order of the planets, beginning with the most remote, is Saturn, Jupiter, Mars, the Sun, Venus, Mercury, the Moon. Now, the day being divided into twenty-four hours, each hour was consecrated to a particular planet, namely: One to Saturn, the following to Jupiter, the third to Mars, and so on according to the above order; and the day received the name of the planet which presided over its first hour. If, then, the first hour of a day was consecrated to Saturn, that planet would also have the 8th, the 15th and the 22nd hours; the 23d would fall to Jupiter, the 24th to Mars, and the 25th or the first hour of the second day would belong to the Sun. In like manner the first hour of the third day would fall to the Moon, the first hour of the fourth to Mars, of the fifth to Mercury, of the sixth to Jupiter and of the seventh to Venus. The cycle being completed, the first hour of the eighth day would again return to Saturn and all the others succeeded in the same order."—*Our Calendar,* pp. 13-14.

[94] Cf. Wilson, *The Romance of the Calendar,* p. 236.

These pagan names were used by the Christians from an early date.[95] The first day of the week eventually became known as the Lord's day, the "Dies Dominica," in honor of Christ's resurrection. Thus modified, the Latin names made their way into the French and Italian languages, with the exception of Saturday.[96]

Whether or not the phases of the moon had any influence on the origin of the week in Egypt or elsewhere, it is certain that this time-unit has no reference to the celestial clock. It is an arbitrary unit, just as is the month under the present system. Furthermore, it stands aloof from the other time-units in a class by itself. There are so many days in a month, and so many days and so many months in a year. A new day and a new month begin with the new year. It is not so with the week. This haughty character keeps its own regular sequence and does not adapt itself to the month and the year. The days, months and years have their own identity, but not the week. Thus the days are numbered from 1 to 31, the months have their proper name and the years have their numerical standing in some era or cycle. The weeks, on the contrary, "follow on in endless succession, each one passing nameless in the abyss of the past."[97] We have the expressions this week, next week, last week, and that is about all.

[95] At the time of St. Augustine the practice was fairly well established to the great sorrow and against the efforts of the Saint. Cf. Leclercq, "Les Jours de la Semaine,"—*Dictionnaire d'Archéologie Chrétienne et de Liturgie* (14 tomes in 27 vols., Paris: Letouzey et Ané, 1924-1939), VII. 2737.

[96] In French one has the following: Dimanche, Lundi, Mardi, Mercredi, Jeudi, Vendredi, (Samedi).—In Italian we have: Domenica, Lunedì, Martedì, Mercoledì, Giovedì, Venerdì, (Sabato). Samedi, as well as Sabato, is derived from "sabbati dies."

The liturgical terminology of the Church does not adhere to the pagan wording for obvious reasons. Portuguese is possibly the only language derived from the Latin which does not use the Roman terms. The days of the week are called "feira," from the Latin "feria," and are likewise enumerated. This practice is due to a great extent to the efforts of St. Martin of Braga (†580). Cf. McKenna, *Paganism and Pagan Survivals in Spain up to the Fall of the Visigoth Kingdom,* The Catholic University of America, Studies in Medieval History, New Series, vol. I, (The Catholic University of America: Washington, D. C., 1938), p. 93.

[97] Philip, *The Calendar,* p. 41.

Chapter II

THE CLOCK

The second time-keeper informs us as to the length of the day and of its subdivisions, the hours, the minutes and the seconds. These minor units are purely conventional and never had any relation with the motion of the celestial bodies and need not here be studied in a special way. The day on the contrary has a most fascinating history. It is and always has been determined by the earth's behavior, and yet the reckoning of the day is now totally different from what it was two thousand years ago.

Our day is an interval of 24 hours reckoned from midnight to midnight.[1] It is based on scientific computattions and observations which were unknown to the ancients. This interval of 24 hours is closely associated with the earth's rotation on its axis, and corresponds, contrary perhaps to popular belief, to a little more than one complete turn.

Article 1—The Primitive Concept of Day

The knowledge of the ancients was limited to the perception of the perpetual recurrence of light and dark. These two periods were considered as distinct. Day ran from dawn to dusk, and night from dusk to dawn. The Roman day consisted of 12 hours which covered the interval of time from the rising to the setting of the sun. The hour was then only a flexible unit and corresponded to the 12th part of any given day. It was longer or shorter according to the seasons. The period of light on December 23, for instance, lasted 8 hours and 54 minutes. The Roman hour for that day was only 44 minutes and 30 seconds long. On the other hand the bright period of June 25 lasted 15 hours and 6 minutes. The hour then ran its course through 75 minutes and 30 seconds.

[1] The word hour (hora, heure, etc.) is derived from the Egyptian god "Horus."—Cf. Antonellus, *De Tempore Legali,* lib. I, cap. III, n. 40.

Lacau gives us the following table of the concordance between that system and our own:

AEQUINOCTIUM			SOLSTITIA	
			Solstitium aestivum	Solstitium hiemale
	seu		seu	seu
	la dies veris et autumni		la dies aestatis	la dies hiemis
Ortus solis;	la hora	—6h	—4h 27m	—7h 33m
	2a "	7h	5 42 30s	8 17 30s
	3a "	8h	6 58	9 2
	4a "	9h	8 13 30	9 46 30
	5a "	10h	9 29	10 31
	6a "	11h	10 44 30	11 15 30
Meridies:	7a "	12h	12	12
	8a "	13h	13 15 30	12 44 30
	9a "	14h	14 31	13 29
	10a "	15h	15 46 30	14 13 30
	11a "	16h	17 2	14 58
	12a "	17h	18 17 30	15 42 30
Occasus solis:		18h	19 33	16 27 [2]

Thus one sees that the only hour with an invariable beginning was the seventh. It always coincided with noon. Strange to say, the English word "noon" comes from *"nona,"* or ninth hour.

In every-day use, however, these various hours were generally relegated to the background and the day was simply divided into four parts:

1—*Mane*—from sunrise or the first hour to the beginning of the third hour.

2—*Ad meridiem*—from the third to the end of the sixth hour.

3—*De meridie*—from the seventh to the beginning of the eleventh hour.

4—*Suprema*—from the eleventh hour to the end of the twelfth hour, or sunset.

The night was likewise divided into four parts called vigils or watches. Each one of these had an average length of three hours. This duration naturally varied with the seasons. The third watch, however, always began at midnight.[3]

[2] *De Tempore* (Romae: Marietti, 1921), p. 37.

[3] Cf. *loc. cit.*

That ancient mode of reckoning is not totally extinct. There is a remainder of it in canon 821, § 1.[4]

A possibly more familiar instance of this reckoning is given us in the Gospels relative to Our Lord's death. One reads in St. Matthew: "Now from the sixth hour there was darkness over the whole earth, until the ninth hour. And about the ninth hour Jesus cried with a loud voice saying: *Eli, Eli lamma sabacthani?* that is, My God, My God, why hast thou forsaken me?"[5]

For the Jews the alternation of light and dark constituted a day, but the reckoning was made from sunset to sunset. There is a mention of this in the first book of the Old Testament: "And he called the light Day, and the darkness Night; and there was evening and morning one day."[6] The Moslems have the same arrangement. According to Wilson: "They have to set their clocks at 12 every sunset and such a day was in use throughout Greece, possibly perpetuated by Turkish rule, during the 19th century."[7]

That system was for a time commonly used in Rome also.[8] Strange to say, some remnants of this form of reckoning are still extant in the Eternal City. Students especially who have spent some time there are fully aware of the meaning of the "Ave Maria" or "Angelus Domini" bell. In America the Angelus is always rung at the same hour each day. It is not so in Rome. The evening Angelus bell is rung at sunset and its time varies accordingly. However, the ringing of the bell does not vary every day as does the setting of the sun, but only when the accumulated difference adds up to 15 minutes, according to the well known table:

TABELLA PEL SUONO DELL'AVE MARIA

a tempo medio astronomico

	ore	q.			ore	q.	
14 Gennaio	17	2	(5:30 P.M.)	1 Agosto	19	3	(7:45 P.M.)

[4] "Missae celebrandae initium ne fiat citius quam una hora ante auroram vel serius quam una hora post meridiem."

[5] XXVII, 46.

[6] Genesis, I, 5.

[7] *The Romance of the Calendar*, p. 208.

[8] Cf. Lacau, *De Tempore*, p. 37.

27 "	17	3	(5:45 P.M.)	11 "	19	2	(7:30 P.M.)
9 Febbraio	18	—	(6:00 P.M.)	21 Agosto	19	1	(7:15 P.M.)
22 "	18	1	(6:15 P.M.)	31 "	19	—	(7:00 P.M.)
7 Marzo	18	2	(6:30 P.M.)	8 Settembre	18	3	(6:45 P.M.)
20 "	18	3	(6:45 P.M.)	16 "	18	2	(6:30 P.M.)
2 Aprile	19	—	(7:00 P.M.)	24 "	18	1	(6:15 P.M.)
15 "	19	1	(7:15 P.M.)	4 Ottobre	18	—	(6:00 P.M.)
28 "	19	2	(7:30 P.M.)	13 "	17	3	(5:45 P.M.)
11 Maggio	19	3	(7:45 P.M.)	22 "	17	2	(5:30 P.M.)
24 "	20	—	(8:00 P.M.)	4 Novembre	17	1	(5:15 P.M.)
11 Giugno	20	1	(8:15 P.M.)	20 "	17	—	(5:00 P.M.)
15 Luglio	20	—	(8:00 P.M.)	28 Dicembre	17	1	(5:15 P.M.)[9]

The ringing of the bell at sunset not only invites the Romans to the recitation of the Angelus. It marks the closing time of many churches, it serves as a regulator for divers ecclesiastic and scholastic functions. Afternoon lectures in the Roman universities, for instance, will not begin at 3 P.M. all year round. They are given one, two or three hours before the Angelus or Ave Maria. The first afternoon session might begin at 3 o'clock in December, but not before 5 o'clock after Easter.

For general use, however, sunrise and sunset have long been discarded as the beginning of day for the simple reason that they vary with the season in any given place except the equator. If one now consider the whole globe one finds that these two phenomena vary with the latitude and that night and day are of equal length only twice a year, on the vernal and autumnal equinoxes. At the equator night and day have exactly twelve hours each throughout the year. The poles on the other hand have a year made up of one night and one day. These last six months. There sunrise and sunset have never had any meaning as the measure of day.[10]

ARTICLE 2—THE MODERN CONCEPT OF DAY

The irregular alternation of light and dark depends on the double motion of the earth, namely, the rotation on its axis and the revolution around the sun along its oblique orbit. In due course

[9] Cf. Lacau, *op. cit.* p. 38.

[10] Cf. Flammarion, *Astronomie Populaire,* p. 36, where various tables are given concerning the duration of the day.

of time scientific developments led man to the conclusion that the earth's rotation alone was a much better time-keeper and that it hardly varied. The sun-dial told him that the sun crossed the meridian at approximately the same time every day irrespective of the moment of dawn and dusk. A day was then associated with one complete spin of the globe on its axis, and had the same value at the poles as at the equator. According to this system there are two natural moments adaptable for use as the beginning of day: noon and midnight. "Noon is the moment when the sun crosses the meridian and is thus highest above the horizon. Midnight is the corresponding period when the sun is lowest below the horizon." [11]

In civil reckonings midnight is nearly everywhere used as the beginning of day for practical reasons. Astronomers, however, at least until a few years ago, were measuring the day from noon to noon. In fact these days are the same.

The problem that must now be solved is the determination of noon and midnight. These two terms simply denote either the beginning or the middle of a day, but they themselves, even though regulated by the rotation of the earth on its axis, have no absolute value. They are relative quantities, and the nature and the length of the days that they mete out to us depends on the starting point that we use for our computation. If the starting point be a star or some fixed point in the heavens we have the sidereal day; if the starting point be the sun, we have the solar day.

A sidereal day may therefore be defined "the interval of time between two consecutive transits of any star over the same celestial meridian." [12] This sidereal day is completed in one absolute rotation of the earth on its axis and takes approximately 23 hours and 56.4 minutes civil clock time. It never varies. Astronomers divide this day into 24 sidereal hours; its time is kept by a sidereal clock. Needless to say, this sidereal time is used only

[11] Wilson, *The Romance of the Calendar*, p. 210.

[12] Newcomb and Holden, *Astronomy*, p. 43. Celestial meridians correspond to the meridians we have often noticed on maps. They are imaginary circles in the heavens, and pass through the north point, zenith, and the south point, nadir.

in scientific computations. There are 366¼ sidereal days in the tropical year.

A solar day is the interval of time during which the sun equally illumines every point of the earth's equatorial circumference. Or if you wish to use a definite starting point, a solar day is simply the interval of time between two successive transits of the sun's center over any given meridian.[13]

That time is called apparent or true sun time and is recorded by the sun-dial.[14] It will be found that it is noon at the same time everywhere along the same meridian. When it is noon in Washington, D. C., for instance, it will also be noon along the 77th meridian (as in Westminister, Md., for instance), but it will already be past noon (P.M.) everywhere east of Washington, as in Boston and New York, and not yet noon (A.M.) west of that point, as in Chicago and San Francisco. The earth's equatorial circumference is divided into 360 degrees which are all reached by the sun in the space of 24 hours. If we divide 360 by 24 we find that one hour corresponds to 15 degrees. One degree is equivalent to one hour or 60 minutes divided by 15, that is, 4 minutes. Therefore when it is noon in Chicago, it is 1:06.5 P.M. in Boston, which is 16⅝ degrees east of Chicago, and 10:50.5 A.M. in Denver, which is 17⅜ degrees west of Chicago.

But, days measured in accordance with true sun time are not equal in length. If we compare the clock and the sun-dial we find that the clock will run ahead of the sun-dial during certain periods of the year, but will lag behind during other periods. True solar time does not therefore fulfill the requirements of a civilized society. It is still used only in the more primitive communities, where the sun-dial is the official time-keeper.

Men in civilized countries soon set this true solar time aside for the use of a modified form called mean solar time. The mean solar day is nothing but the average of all the true solar days of the year. This time is regular and can be kept by ordinary clocks

[13] Cf. Godfray, *A Treatise on Astronomy*, p. 151.

[14] Cf. Schwegler, "Sun Time Simplified,"—*ER,* CI (1939), 3, where a few considerations are given concerning the words apparent and true sun time.

and watches. It is often called civil time; its day always corresponds to 24 of our watch hours. Astronomers reckon this time from the motion of a fictitious sun which is imagined to move uniformly in the equator, but that is of little importance to us.

The difference between true solar and mean solar time is attributed to two major causes, namely, 1—the unequal motion of the earth in its annual revolution around the sun, arising from the eccentricity of the earth's orbit, and 2—the obliquity of the ecliptic.[15]

1—Eccentricity of the earth's orbit—

The first thing to be borne in mind is the fact that while the earth whirls around the sun there is a relation between the speed of the earth and the length of the day. It has already been noted that the length of the sidereal day is 23 hours and 56 minutes, and that it corresponds exactly to one complete turn of our sphere on its axis. If for some reason or other the earth stopped moving through space, our solar days also would last only 23 hours and 56 minutes. On the other hand, if it moved faster than it now does the days would be longer than 24 hours.

As things now stand the earth rotates 366¼ times a year, but that give us only 365¼ solar days. In other words, one of the earth's rotations is offset by the apparent motion of the sun in the opposite direction. In one sidereal day the earth rotates exactly 360° and in a mean solar day 360° 59' 8." At the same time the sun is apparently moving west among the stars at the average speed of 59' 8". In fact the actual speed will vary from 0° 57' 11.5" to 1° 1' 9.9".[16]

The reason for this variation is due to the eccentricity of the earth's orbit around the sun. The earth indeed does not follow a perfectly spherical course as the ancients believed; its distance from the sun varies slightly, the point when it is closest to the sun being called the perihelion, and the point when it is remotest

[15] Cf. Newcomb and Holden, *Astronomy*, p. 188.—Other causes also affect the equation of time, but never more than a very few seconds. Cf. Young, *Elements of Astronomy*, p. 21.

[16] Cf. Godfray, *A Treatise on Astronomy*, p. 142.

from the sun being called the aphelion. Perihelion now occurs on January 2.

According to a well known law the earth moves faster when closer to the sun than it does when farther away. When the terrestrial sphere is in the surroundings of the perihelion its velocity will exceed its mean value and the true solar days will be longer than the mean solar days. Vice versa, when the earth approaches the aphelion its velocity will not equal its mean value and the true solar days will be shorter than the mean solar days. The difference in time due to this cause alone—the eccentricity of the earth's orbit—amounts to eight minutes.[17]

2—Obliquity of the ecliptic—

This second cause accounts for a difference of about ten minutes between true sun time and mean sun time.[18]

The apparent path of the sun around the earth lies in a plane inclined 23° 27′ to the plane of our equator. That inclination alone does not cause an actual variation in the absolute speed of the earth's motion, but it so affects the relation of the earth to the sun as far as illumination is concerned, that its effect works much in the same way as the previous cause. Abstracting from the eccentricity of the earth's orbit one finds that, according to the obliquity of the ecliptic alone, there are four days of equal duration during the year, namely, the two equinoxes and the two solstices.

If one now combines the two causes one finds that true sun time and mean sun time correspond to each other on April 15, June 15, August 31, and December 24.[19]

The variation between true sun time and mean sun time in one single day is sometimes as high as 22 seconds. From December 24 to February 11 days measured in accordance with true sun time get longer and longer, so that the accrued variation mounts up to 14.5 minutes on the latter date. From February 11 to May 14 the reverse process is true. On the latter date the sun-

[17] Cf. Young, *Elements of Astronomy*, p. 81.

[18] *Loc. cit.*

[19] Cf. Godfray, *op. cit.*, p. 155.

dial would register noon 3.9 minutes earlier than a clock which is set in accordance with mean sun time on any given meridian.

The difference of time between true sun time and mean sun time at any instant is called the equation of time. "It is usually considered as the correction to be applied to the former to obtain the latter, and is therefore called positive when mean noon precedes true noon, and vice versa." [20]

The following table gives an idea of the equation of time for the whole year:

+14.5 minutes about Feb. 11
− 3.9 minutes about May 14
+ 6.2 minutes about July 25
−16.3 minutes about Nov. 1.[21]

> The time-keepers formerly used were so imperfect that these inequalities in the solar day were nearly lost in the necessary irregularities of the rate of the clock. All clocks were therefore set by the sun as often as was found necessary or convenient. But during the last century it was found by astronomers that the use of units of time varying in this way led to much inconvenience; they therefore substituted mean time for solar or apparent [true] time.[22]

According to Flammarion public clocks in France began to give mean time in 1816.[23] But even a century earlier watch-makers were fully aware that their clocks were more uniform time-keepers than the sun.[24]

Thus far local time alone, both true and mean, has been considered. Local time is determined by the passage of the real or the fictitious sun over the meridian of any given community. Until recent years, when the means of communication were rather

[20] Cf. Godfray, *op. cit.* pp. 152-153.

[21] Cf. Godfray, *op. cit.*, p. 156.—A complete table is found in Appendix I (pp. 257-264) of this work.

[22] Newcomb and Holden, *Astronomy*, p. 189.

[23] *Astronomie Populaire*, pp. 24-25.

[24] Cf. Flammarion (*op. cit.*, p. 25): "Dès le temps de Louis XIV, la communauté des horlogers de Paris avait pris pour armoirie une pendule avec cette orgueilleuse devise: *Solis mendaces arguit horas:* Elle prouve que les heures du soleil sont menteuses."

primitive, each city was more or less independent of the others and could follow the mean time of its own meridian irrespective of what other cities were doing. Today such a situation would be intolerable.

"The confusion of clocks," writes Wilson, surveyed in retrospect, is scarcely credible. At railway stations in Paris, the clocks outside the building were five minutes ahead of clocks inside. The reason is stated to have been that railway time was according to Rouen where railways were managed, and Rouen is five minutes ahead of Paris." [25]

That state of confusion finally disappeared toward the end of the last century when some sort of regional, national or zone time was introduced. At that time the European countries imposed one and the same time for their respective territories. That time was usually the mean time of the capital. That form of regional or national time is still in use in the Netherlands.

In the United States the problem was solved in a different way. Regional time came into use without being imposed by the government. Private initiative, led by the railroads, divided the country into four regions or time zones centered roughly about the 75th, 90th, 105th, and 120th meridians. That regional time came into use November 18, 1883.[26]

In the meantime Sir Sandford Fleming had proposed the Standard Time plan as we know it today. The world was to be divided into 24 even zones of 15 degrees longitude and every-

[25] *The Romance of the Calendar,* p. 305.—"A hundred years ago, little as we now realize it, there was thus a time of the day, even in a small country like England, that varied everywhere as you travelled by train or coach. A time-table issued on July 30, 1841, announced:

"LONDON TIME is kept at all the Stations of the Railway, which is about 4 minutes earlier than READING time, 5½ minutes before STEVENTON time; 7½ minutes before CIRENCESTER time; 8 minutes before CHIPPENHAM time; 11 minutes before BATH and BRISTOL time; and 14 minutes before BRIDGEWATER time."—*Loc. cit.*

[26] Cf. Fleming, "Time Reckoning for the 20th Century,"—*Annual Report of the Board of Regents of the Smithsonian Institution,* 1886, Part I, p. 357; Riegel, "Standard Time in the United States,"—*The American Historical Review,* XXXIII (1927-1928), pp. 84-89.

where in that zone the same time was to be kept. This was adopted in principle during a conference held in Washington, D. C., October, 1884, in which 25 nationalities were represented. Greenwich was considered the Prime Meridian or 0° longtitude. It was the center of a zone of 15° so that all the regions between 7½ longitude west of Greenwich and 7½ longitude east of Greenwich were to use Greenwich mean time. Thus the 15th meridian was the center of another time belt that took in the regions between the 7½ and the 22½ degrees east or west of Greenwich. In that case the time was one hour faster or slower than Greenwich time. And so on throughout the world.[27]

The 180th meridian which is found in the mid-Pacific, has a special importance. In relation to the noon of Greenwich time it indicates midnight, and has been called the Calendar Line or the International Date Line. When the sun crosses the 180th meridian it will be 12 o'clock noon of March 24 west of that meridian, but 12 o'clock noon of March 25 east of the same meridian.

Cardenas writes:

> Let us suppose that we board an airplane in New York, at Sunday noon, and travel westward with exactly the same speed as the sun, keeping it always over our heads. Obviously it is noon in every place we visit. If we ask the day of the week at every point of the journey, we will be told that it is Sunday in all points of the United States, and even as far west as Hawaii. This cannot hold true indefinitely, for when we finish our flight around the world, we shall have been gone for 24 hours, and consequently we shall be told in New York upon our return that it is Monday noon.[28]
>
> Everywhere in Asia and Europe we should have been told that it was Monday noon. Yet it has always been the

[27] Cf. Fleming, *op. cit.* pp. 345-366. Cf. also *Standard Time Throughout the World,* Circular of the National Bureau of Standards C 406 (Washington: United States Government Printing Office, 1935), p. 3, where the reader is able to find the dates when the Standard Time plan was adopted by the various nations.—"Strange to recall, Congress only legalized these zones in 1918, 40 years after Sandford Fleming's proposal,"—Wilson, *The Romance of the Calendar,* p. 307.

[28] "Midnight and Canon Law,"—*ER,* CI (139), 411.

> same day to us, and there must have been some place in the journey where we were told for the first time that it was Monday instead of Sunday.[29]

This place is the International Date Line that has already been mentioned: it just about follows the 180th meridian.

That meridian in some instances runs through a city, a group of islands, etc. It is obvious that these communities must have the same time. The early settlers in these parts of the Pacific undoubtedly adopted the day commonly used in the land of their origin. Thus the settlers that came around the Cape of Good Hope used the Asiatic day. The International Date Line consequently deviates to the west to include the Fiji and Tonga Islands with Australasia. The extremity of Siberia is also west of 180, but nevertheless follows the Asiatic day. Similarly the International Date Line deviates in the opposite direction so as to give the Aleutian Islands the American day.[30]

Alaska presents an interesting case. Before 1867 the day was the one the Russians had brought over from Siberia. When American Catholics first settled in Alaska the Russian Christians were celebrating their weekly services on Saturday according to the American way of reckoning. On the other hand the Russians thought that the Catholics were fulfilling their weekly Mass obligation on Monday. This awry situation was definitely settled when the United States purchased Alaska in 1867. One day was repeated.[31]

The International Date Line gives rise to many problems to the Far East traveler. An effort will be made to examine them in the commentary.[32]

These remarks bring to a close the first part of this dissertation. Thus far time has been considered in itself, without reference to it legal aspect. The nature of the civil calendar was explained, and a few considerations were given concerning the nature and

[29] *Loc. cit.*
[30] Cf. *Standard Time Throughout the World,* p. 8.
[31] Cf. Cardenas, "Midnight and Canon Law,"—*ER,* CI (1939), 411.
[32] Cf. *infra,* pp. 186-188.

length of the various time-units now in use. The moment has now come to study the subtle element of time from the juridical point of view.

PART II—THE JURIDICAL RECKONING OF TIME

Section I—Historical Synopsis

Chapter III

PRELIMINARY OBSERVATIONS

Article 1—General Remarks

The time-units that have been considered in the first part of this work, namely, the year, the month, the week and the day, have practically the same meaning in every science. Slight differences, however, occasionally creep in. Thus the Roman civil day (*dies*) consisted of twelve hours reckoned from dawn to dusk. This period was opposed to the night (*nox*) which ran from dusk to dawn.[1] On the other hand the Roman legal day was reckoned from midnight to midnight, as shall be shown in the following chapter. Likewise the civil year of the Jews differed from their sacred year.[2]

The Church did not escape the general law according to which all institutions here below are at one time or other affected by their surroundings. Jerusalem and Rome have played a predominant part in the development of the Church. It is not therefore surprising to find their imprint on the institute of time as well as on many others.

Today there are two distinct forms of time-reckoning in the Church, one being exclusively used in liturgical affairs, the other in disciplinary matters. The former is to a great extent dependent on the Jewish mode of computation. The week, as we know it, is fundamentally a Jewish institute, and the reckoning of Easter with all the feasts that depend on it is of the same origin. The computation of the day from vespers to vespers likewise bears the imprint of the Hebrew ceremonial laws.[3]

[1] Cf. *supra*, pp. 38-39.

[2] Cf. *infra*, pp. 52-53.

[3] This particular form of reckoning was not in former years confined to purely liturgical affairs. For many centuries it was practically the only

The influence of the Romans on canonical time-reckoning is still greater. They gave us at least the framework of the Gregorian calendar. The direct influence of the Roman legal institutes has been felt to the present day in the compuation of the year and the month. In purely disciplinary matters the Roman reckoning of the day from midnight to midnight has had, for well nigh a thousand years, an ever increasing ascendency.

It is therefore imperative that a few words be said about these two diverse forms of time-reckoning, the Jewish, and especially the Roman, before the history of this institute is studied in Canon Law itself.

Article 2—The Jewish Reckoning

The vicissitudes and the migrations of God's chosen people seem to have played an important rôle in their calendar making. Thus one finds that at the time of Moses the Hebrews used a calendar very similar to that of the Egyptians. It consisted of twelve solar months, the first eleven of which contained thirty days. The twelfth month had thirty-five days.[4]

The lunar month, which alone is of interest here, was introduced after Alexander the Great's time.[5] The Scriptures themselves show us the importance of the moon as a time-keeper:

> And the moon in all in her season, is for a declaration of times and a sign of the world. From the moon is the sign of the festival day, a light that decreaseth in her perfection. The month is called after her name, increasing wonderfully in her perfection.[6]

This lunar month was the only one used, but it served in a twofold capacity for the simple reason that the Hebrews reckoned

known reckoning of the day, if one excepts strictly juridical computations. Cf. *infra*, pp. 90-93.

[4] Cf. Calmet, *Dictionnairc Historique, Archéologique, Philologique, Chronologique, Géographique et Littéral de la Bible* (4. ed., 4 tomes, Paris, 1846), s.v. *Mois*.—This work will hereafter be cited simply as *Dictionnaire de la Bible*. Cf. *supra*, pp. 8-9, concerning the Egyptian calendar.

[5] Cf. Calmet, *loc. cit.*

[6] Ecclesiasticus, XLIII, 6-8.

the beginning of the year at different times according to the nature of the matter involved. There were, among others, the civil year and the sacred year, the former beginning with the month Tizri, the latter with the month Nisan.[7]

The actual length of the lunar month is 29 days, 12 hours, 44 minutes and 2 seconds.[8] The Jews, however, roughly considered its length to be 29 days and 12 hours. Their calendar was so constructed as to have months of 30 days alternating with months of 29 days, the 30 day months being called *full,* the others, *empty.*[9]

The new moon meant the beginning of a new month, and was called *Neomenia.* This phenomenon was not determined by scien-

[7] It might be of interest to give here the names of the Jewish months and their order for both the sacred and the civil year:

SACRED YEAR

1—Nisan—corresponding approximately to	March
2—Jiar	April
3—Sivan (Sivam)	May
3—Thamnuz	June
5—Ab	July
6—Elul	August
7—Tizri	September
8—Marschehhon (Marsheran)	October
9—Casleu	November
10—Thebet	December
11—Sebat or Sabat	January
12—Adar	February

CIVIL YEAR

1—Tizri	September
2—Marsheran (Marschehhon)	October
3—Casleu	November
4—Thebet	December
5—Sebat	January
6—Adar	February
7—Nisan	March
8—Jiar	April
9—Sivam (Sivan)	May
10—Thamnuz	June
11—Ab	July
12—Elul	August.

Cf. Calmet, *Dictionnaire de la Bible,* s.v. *Mois.*

[8] Cf. Godfray, *A Treatise on Astronomy,* p. 229.

[9] Cf. Calmet, *Dictionnaire de la Bible,* s.v. *Mois;* Dionysius Exiguus, *Epistolae Duae de Ratione Paschae,—MPL,* LXVII, 21.

tific methods. The visibility of the new moon was alone taken into consideration. Men kept a vigilant eye on the high places in order to warn the Sanhedrin within the shortest possible delay. The sound of trumpets announced the new month.[10]

Twelve of these months gave only 354 days, while the length of solar year is 365.2422 days. Intercalations had to be used to harmonize the lunar year with the solar cycle.[11] The thirteenth or intercalated month was called *Ne-Adar,* or *Second-Adar,* and was wedged in between Adar and Nisan, according to regulations emanating from the Sanhedrin.[12] In due course of time these intercalations became regulated according to a cycle of 19 years, much similar to the Metonic cycle.[13]

During the course of this cycle there were 235 lunations which accounted for 6939 days, 16 hours and 31 minutes, whereas the true length of these 19 years was 6939 days, 14 hours and 27 minutes.[14] The difference between the 235 lunar months and the 19 true solar years was hardly more than two hours.[15]

[10] Cf. Calmet, *loc. cit.*

[11] Cf. *supra,* pp. 9-11.

[12] Cf. Calmet, *loc. cit.*

[13] Cf. *supra,* pp. 9-10. This cycle was divided into two series or parts. The first of these consisted of 8 years and was called *ogdoas,* while the other was made up of 11 years and called *hendecas.* Within this cycle the years were arranged in the following order:

Common years of 12 months—
Ogdoas —1—2—4—5—7
Hendecas —1—2—4—5—7—8—10
Embolismal years of 13 months—
Ogdoas —3—6—8
Hendecas —3—6—9—11.

Cf. Tagle, *De Temporis Supputatione* (Washington, D. C., 1932), pp. 17-19; Venerable Bede, *De Ratione Computi,—MPL,* XC, 593-598; Rabanus Maurus, *De Universo,—MPL,* CXI, 125-127; Honorius Augustodunensis, *De Imagine Mundi,—MPL,* CLXXII, 155.

[14] Cf. *supra,* p. 9.

[15] Cf. St. Cyprian: "Hac itaque ratione non sua sed Dei sapientia instructi, Hebraei circa cursum lunarem, juxta regulam primam Graecorum, more Aegyptiorum, et non secundum epactas lunares, non potuerunt errare."—*De Pascha Computus,—MPL,* IV, 949. CF. also Tagle, *De Temporis Supputatione,* pp. 16-20.

Another Jewish institute that has profoundly influenced the reckoning of time in the Church is the week. In Calmet one reads that the Hebrews have always used the week as a time-unit to commemorate the creation of the world which was accomplished in seven days.[16] The only day of the week to have a proper name was the Sabbath, the others having only a numerical value with reference to the Sabbath.[17]

The reckoning of the day was from sunset to sunset, and this computation originally applied to the civil as well as to the sacred day.[18] "It is a sabbath of rest, and you shall afflict your souls *beginning on* the ninth day of the month, from evening until evening you shall celebrate your sabbaths." [19]

The Jewish Pasch was always celebrated on the fourteenth of Nisan. The intercalations were so made that the fourteenth day of the month of Nisan (full moon) coincided with the vernal equinox or immediately followed it. In other words the Pasch fell on the first full moon after the vernal equinox.[20]

The rules for the determination of Easter in the Church are substantially the same as those of the Jews. Slight modifications, however, gave rise to century long controversies. Since this question has not a direct bearing on the interpretation of canons 31-35, it should be sufficient to present the matter as summarized by Ojetti:

> Una ex praecipuis his legibus liturgicis ea est, quae refertur ad celebrationem paschatis, et in Ecclesia non

[16] "Les Hébreux ont toujours compté par semaines, en mémoire de la création du monde, qui se fit en sept jours."—*Dictionnaire de la Bible,* s.v. *An.*

[17] Cf. Venerable Bede: "Quae populo Dei hebdomada ita computabatur antiquitus: prima Sabbati, vel una Sabbati, sive Sabbatorum: secunda Sabbati, tertia Sabbati, . . . septima Sabbati, vel Sabbatum."—*De Temporum Ratione,—MPL,* XC, 327.

[18] Cf. Calmet, *op. cit.,* s.v. *Jour.*—At the time of Our Lord the civil reckoning of the day was according to the Roman style. Thus one reads in the Gospel according to St. John (XI, 9): "Jesus answered: Are there not twelve hours of the day?"—Cf. Mark, VI, 48; XIII, 35; XV, 34; Matthew, XIV, 25; XX, 3-5; Calmet, *op. cit.,* s.v. *Heures.*

[19] Leviticus, XXIII, 32. Cf. Calmet, *loc. cit.*

[20] Cf. Calmet, *Dictionnaire de la Bible,* s.v. *Mois.*

parvas suscitavit aliquando difficultates et dissidia. Quaestio notissima est. Asiatici differebant a Romanis in assignando die mortis Christi (πάσχα σταυρώσιμον) et consequenter die resurrectionis (πάσχα ἀναστάσιμον). Illi diem mortis Christi semper celebrabant die 14 (ιδ') mensis Nisan, quicumque esset dies hebdomadae, et duos post dies, die 16 mensis, celebrabant pascha. Romae dies mortis, semper die Veneris celebratur, vel 14 die Nisan, vel die Veneris sequente (post ιδ'): consequenter sollemnia resurrectionis semper die dominica celebrabantur. Brevi, Romae dies hebdomadae, in Asia dies mensis designabat πάσχα ἀναστάσιμον. Inde etiam variatio quaedam in disciplina ieiunii. Omnes maxime afficiebantur in favorem suae consuetudinis appellantes, Asiani quidem ad Philippum et Ioannem Apostolos, Romani ad Petrum et Paulum. Quousque varietas fuit in disciplina et territorium dogmatis non invasit, res non nimis graves fuerunt: sed cito exsurrexerunt haeretici *Quarto-decimani*, qui dixerunt pascha celebrandum ipsa die 14 Nisan simul cum Iudaeis et secundum eorum morem, quia lex iudaica vim suam non amisisset. *Cfr.* Hergenroether-Kirsch, I, 14 § 2. Exorto errore Victor Papa iniunxit ut haberentur synodi, quae morem romanum reciperent, et ita factum est, quia omnes proclamarunt pascha celebrandum esse die dominica, exceptis paucissimis, Polycrate ep. Ephesino et nonnullis episcopis ei adhaerentibus. Eus. H. E. V. 23; Brück, *Storia eccl.*, 37.—Alia fuit quaestio, num dies 14 Nisan poni deberet ante vel post aequinoctium vernum, diversis diversa sentientibus. Hinc *Protopaschitae*, qui volebant pascha celebrandum semper post plenilunium, quod sequebatur aequinoctium vernum; quare non raro statuebant ιδ' (14 Nisan) ante principium veris. Contra alii fere omnes posuerunt ιδ' semper post aequinoctium et celebrarunt pascha prima die dominica post plenilunium incoeptum, vere iam ingresso. *Cfr.* Brück, l. c.; Hergenroether-Kirsch, l.c. Synodus Arelatensis a. 314 statuit pascha celebrandum esse ubique eadem die eodemque tempore. Pontifex Romanus debebat ubique promulgare diem certum celebrationis. Leo M. *Ep.* 121 al. 94.[21]

[21] *Normae Generales*, pp. 192-193, note. Cf. Betten, *From Many Centuries* (New York: Kenedy, 1938), pp. 210-231; *Cath. Hist. Review*, XIV (1928-1929), 485-499.

According to the present discipline, therefore, Easter is always celebrated on a Sunday, the first after the first full moon after the vernal equinox as explained above. The Jewish Pasch was cccasionally celebrated on the day itself of the full moon (14 Nisan). Under the present discipline, however, that day is excluded. If the full moon falls on a Sunday, Easter is celebrated a week later. The reason for this is that the Church does not wish to celebrate this feast on the same day as the Jews.[22]

[22] Cf. *Breviarium Romanum*: "Ne cum Judaeis conveniamus, si forte dies xiv Lunae caderet in diem Dominicum."—*De anno et ejus partibus, De festis mobilibus.* For more details concerning the Easter controversy the reader is referred to Hefele-Leclercq, *Histoire des Conciles,* Tome I, Part I, #37—Solution de la question pascale, pp. 450-477.—For a short commentary on the various time-units such as epacts, Dominical letters, etc., mentioned in the rubric *De anno et ejus partibus* of the Roman Breviary, cf. Phillips, "The Year and its Parts,"—*ER,* XCIII (1935), 1-3.

CHAPTER IV

ROMAN LAW

ARTICLE 1—THE MEASURING OF TIME

The various time-units used in the Roman legal system can be examined from two distinct angles. The first considers these time-units in themselves and merely gives their absolute, theoretical value or length. Thus this consideration tells us that the Roman legal year consisted of 365 days, and that the month had 30 days. The second angle, which is the more complicated, examines the above mentioned units in concrete cases. It determines how the 30 days that compose the month are actually going to be reckoned, whether or not the first day is counted, and whether or not the last day is held as complete even though it be only begun.[1]

Day—The Roman legal day consisted of 24 hours which were reckoned from midnight to midnight.[2]

Month—The ordinary length of the legal month was 30 days.[3] However, the duration of this time-unit was occasionally deter-

[1] Cf. Van Hove, *Commentarium Lovaniense in Codicem Iuris Canonici*, Vol. I, Tomus III, *De Consuetudine—De Temporis Suppuatione* (Mechliniae: Dessain, 1933), n. 271; D'Angelo, *Ius Digestorum* (2 vols., Romae: Athenaeum Pontificii Seminarii Romani ad "S. Apollinaris," 1927), I, n. 839.—Van Hove's work will hereafter be cited as *De Temporis Supputatione.*

[2] "More romano dies a media nocte incipit et sequentis noctis media parte finitur. Itaque quidquid in his viginti quattuor horis, id est duabus dimidiatis noctibus et luce media, actum est, perinde est quasi quavis hora lucis actum esset."—D. (2, 12) 8. Cf. Maynz, *Cours de Droit Romain* (3. ed., 3 vols., Brussels, 1870), I, #125, 1.

[3] "Quando altera pars dicit habere se quid proferre debeat, iudex partem, quae dilatione utitur, intra triginta dies post alterius partis renuntiationem id quod voluerit proponere cogat. Quod si non fecerit, tunc alium mensem iudex largiatur."—Nov. 115, c.2. Cf. also D. (21, 1) 28, 31 #22, 38; D. (48, 5) 12 [11] #6, 29 #5; C. (6, 30) 22 ##2-11.

mined in a different way. One instance gives a total of 93 days for 3 months.[4]

Year—This unit was composed of 365 days.[5]

Week—This unit is practically non existent in Roman Law. Implicit references to it are, however, occasionally found in the sources.[6]

Article 2—The Reckoning of Time

§ 1—General Remarks

Time in Roman Law was not reckoned morally—that is to say, in a rather loose way so as to neglect short intervals—but physically. Everything was taken into consideration.[7]

The actual compuatation of time depended on the manner in which it was granted. One finds in the Roman legal sources two distinct methods of conceding or granting a period of time. The

[4] "Trium mensium spatiis, id est, nonaginta tribus diebus."—C. (7,63) 5 pr. Cf. Maynz, *Cours de Droit Romain,* I, #125, note 3. Other texts seem to imply a somewhat similar reckoning. Thus one reads: "Ubi lex duorum mensium fecit mentionem, et qui sexagesimo et primo die venerit, audiendus est."—D. (50, 17) 101. Authors are not of the same opinion concerning the interpretation of this fragment. For more details cf. D'Angelo, *Ius Digestorum,* I, n. 836, b; Maynz, *op. cit.,* I, #125, note 9.

[5] "Stichus si heredi meo anno servierit liber esto: quaerendum est, annus quomodo accipi debeat, an qui ex continuis diebus trecentis sexaginta quinque constet, an quibuslibet. Sed superius magis intelligendum Pomponius scribit." D. (40,7) 4 #5. Cf. D. (9, 2) 51 #2.—The question of the leap year is treated pp. 65-66 *infra.* The length of the year and the month as of 365 and 30 days respectively came into use with the Julian calendar. For the duration of time before 46 B.C., Van Hove (*De Temporis Supputatione,* n. 271, note 3) refers the reader to E. Brinckmeier, *Praktisches Handbuch der historischen Chronologie,* Berlin, 1882, pp. 23-31; 81-86.

[6] Cf., e. g., C. (3, 12) 6 [7] #4: "In eadem observatione numeramus et dies solis, quos dominicos rite dixere maiores, qui repetito in se calculo revolvuntur."

[7] "In usucapione ita servatur, ut, etiamsi minimo momento novissimi diei possessa sit res, nihilo minus repleatur usucapio, nec totus dies exigitur ad explendum constitutum tempus."—D. (44, 3) 15. The same

first of these mentions the precise day on which a certain act can or must be posited. It also determines a certain day as the *terminus ad quem* of an indefinite amount of time. The second method grants a certain number of days, months or years, to be reckoned from a definite starting point or *terminus a quo*.

I— In the first case, for instance, an individual might be obliged to pay a debt July 2nd, or July 11th. These days are given as the ultimate date for the debtor to solve his obligation. In such cases, if the day alone is mentioned—and such is the rule —without any reference to the hour of the day, a full day is to be counted and the time does not expire before midnight of the above mentioned dates.[8]

If a certain day is mentioned, but without any specific determination as to the year or month, such as *the first of March* without any specified year, then the first recurrence of that date is the accepted one in Roman Law.[9]

II— In the second case the period of time involved is determined by the indication of the extension of time that must lapse from a given moment. Such is the reckoning used when a certain sum of money must be paid "within a month" or "within

conclusion is to be drawn from the works of responsible authors. Toso (*Commentaria Minora,* I, 104) thus quotes the following words of Cicero: "Quae quoniam in temporum inclinationibus saepe parvis posita sunt, omnia momenta observabimus, neque ullum praetermittemus tui juvandi et levandi locum." *Ad Familiares,* 6, 10.

[8] Cf. Windscheid, *Diritto delle Pandette, Traduzione dei Professori Carlo Fadda e Paolo Emilio Bensa* (2. ed., 5 vols., Torino: Unione Tipografico-Editrice Torinese, 1925-1926), I, #103, note 3: "Se si è stabilito un usufrutto, 'fino al 31 dicembre 1877,' nel dubbio deve intendersi fino al 31 dicembre *inclusivo.* Egualmente si interpretava presso i Romani la preposizione *intra*: 'Intra Kalendas etiam ipsae Kalendae sunt.' 1 #9 D. 38, 9." Hereafter this work will be cited simply as *Pandette.*

[9] Cf. D. (45, 1) 41; Ojetti, *Synopsis Rerum Moralium et Iuris Pontificii* (Romae, 1899) s,v. *Tempus.* This work will hereafter be referred to as *Synopsis.* Cf. also D. (45, 1) 42: "Qui 'hoc anno' aut 'hoc mense' dari stipulatus sit, nisi omnibus partibus praeteritis anni vel mensis non recte petet."

two years," etc. The reckoning of the 21 years of age required for one's majority is another example of this form of computation. This second form of time-reckoning is occasionally complicated, since it determines whether or not any given amount of time is to be computed according to the *natural,* or according to the *civil* form of reckoning.

§ 2—The Natural Reckoning

The natural reckoning (*supputatio naturalis*) considers the day as consisting of 24 hours, but does not reckon them from midnight to midnight. The starting point of the day is determined by an event of some kind such as one's birth, the signing of a contract or of a note, etc., The second day according to this form of reckoning begins exactly 24 hours after the event. This computation is also called supputatio *a momento ad momentum,* reckoning from moment to moment.[10]

§ 3—The Civil Reckoning in General

The civil reckoning (*supputatio civilis*) considers the day as an indivisible unity consistng of 24 hours computed from midnight to midnight. Portions or fractions of a day are not considered here. In other words these legal days correspond exactly to the calendar days.[11]

[10] "Minorem autem viginti quinque annis natu videndum, an etiam die natalis sui adhuc dicimus ante horam qua natus est, ut si captus sit restituatur? Et cum nondum compleverit, ita erit dicendum, ut a momento in momentum tempus spectetur."—D. (4, 4) 3 #3. Cf. Ojetti, *Synopsis,* s.v. *Tempus.* Vinnius thus comments upon this fragment: "Curatio non finitur nisi anno vicesimo quinto exacto et completo. Hoc enim utile minori visum est, ut tempus a momento in momentum spectatur; nam si postremo vicesimiquinti anni die coepto aut nondum completo laesus sit, poterit adhuc in integrum restitui."—*Institutionum Imperialium Commentarius* (Venetiis, 1804), lib. I, tit. XXIII, n. 4.—This *supputatio naturalis* of the day is not to be confounded with the *dies naturalis.* This natural day is not a legal entity; it simply refers to the common everyday concept of day from dawn to dusk.

[11] "Annum civiliter, non ad momenta temporum, sed ad dies numeramus."—D. (50, 16) 134. Cf. D. (2, 12) 8.—The expression *supputatio*

According to the modern authors the civil reckoning is the rule in Roman Law. The natural form of computation is exceptional and is found in a few instances only. According to Van Hove it is used in the *restitutio in integrum* and the *fatalia ad appellandum.*[12]

Windscheid is even more categorical. He states that the natural computation is altogether exceptional and that it is found in only one instance.[13] He goes on to say, however, that the older commentators of Roman Law considered the natural reckoning as the rule. Maynz, for instance, is of that opinion.[14]

It is not surprising to find that the natural reckoning was little used. The difficulty of determining the precise moment of any event, signature or birth, led to many misunderstandings. That accounts for the fact that this form of reckoning is little used today.[15]

The consideration of the civil form of reckoning as the rule

naturalis is not as such found in the sources, but certainly is in conformity with the Roman terminology. In many other institutes the word *naturalis* is opposed to *civilis*. Cf. Windscheid, Pandette, I, #103, note 6; #148, note 12.

[12] *De Temporis Supputatione,* n. 271, 2.—Cf. D. (4, 4) 3 #3; Nov. 23, c. 1; D'Angelo, *Ius Digestorum,* I, n. 840

[13] "Questa regola, a dir vero, non è in alcun luogo espressa in via di principio, ma in buon numero di passi si procede a stregua di essa. L. 5. D. *qui test.* 28, 1,—L. 1. D. *de man.* 40, 1; L. 1, 6, 7, D. *de usurp.* 41, 3; L. 134, D. *de V. S.* 50, 16. Solo in un unico caso, eccezionalmente si applica il computo naturale. L. 3 #3, D. *de min.* 4, 4 . . . "—*Pandette,* I, #103, note 7.

[14] *Cours de Droit Romain, I,* #125, 4.—Cf. Windscheid, *Pandette,* I, #103, note 7, where mention is made of a few of the older commentators who held that the natural reckoning was the rule in Roman Law.

[15] "Haec supputatio naturalis de momento ad momentum valde difficilis est, molestiis et scrupulis obnoxia; nam praesupponit momentum illius facti, a quo tempus supputandum est indubie constare et certe memoria teneri. Unde etiam a iuribus modernis passim exulat."—Wernz-Vidal, *Ius Canonicum,* Tomus I, *Normae Generales* (Romae: Apud Aedes Universitatis Gregorianae, 1938), n. 242, II. This work will hereafter be referred to as *Normae Generales.* Cf. Windscheid, *Panette,* I, #103, note 9; Maynz, *Cours de Droit Romain,* I, #125, 3.

paves the way for other questions. Is the starting point (*terminus a quo*) considered as the first legal day, or must one begin to count only with the following day? Is the last day (*terminus ad quem*) held as complete as soon as it begins?

§4—The Question of the *Terminus a Quo* and the *Terminus ad Quem*

The common opinion of today holds that the *terminus a quo* is counted, or held as a complete day, irrespective of the moment of the day when the juridical act is posited.[16]

The problem of the *dies ad quem* is much more complicated. According to D'Angelo the following theories have at some time or other been proposed.[17]

I—A certain number of authors hold that the general rule is that the last day must always be completed (*Postremus totus dies semper completus esse debet.*) In certain exceptional cases that day is counted as complete even though it has only begun (*Dies coeptus pro exacto* [completo] habetur).

II—Others simply reverse the procedure and contend that the general rule demands that the last day be counted even though only begun.

III—Others, on the other hand, distinguish:

a—If something is to be acquired, then the rule is *dies coeptus*. The rule *postremus totus* applies to losses.[18]

[16] "Dies *termini a quo* pro completo habetur, etiam si actus iuridicus quavis hora diei fuerit positus, seu dies a quo in termino computatur."—Van Hove, *De Temporis Supputatione*, n. 271, 2. Cf. Wernz-Vidal, *Normae Generales*, n. 242, IV.—The fragment D. (41, 3) 7 seems to substantiate this statement. It reads thus: "Ideoque qui hora sexta diei kalendarum Ianuariarum possidere coepit, hora sexta noctis pridie kalendas Ianuarias implet usucapionem."—One must however suppose that this *hora sexta* or midnight marks the beginning and not the end of December 31 (*pridie kalendas*). The wording is of itself not too clear. Cf. Windscheid, *Pandette*, I, #103, 2, a. The question is thoroughly examined in note 12. Cf. also D'Angelo, *Ius Digestorum*, I, n. 841, d.

[17] *Ius Digestorum*, I, n. 841, I, a, b, c.

[18] This, for example, is the opinion held by Wernz-Vidal (*Normae Generales*, n. 242, IV).

b—The rule *dies coeptus* applies in favorable matters. The other holds good for odious affairs.[19]

D'Angelo expresses his own opinion when he writes that the rule *dies coeptus* 1—does not constitute an exceptional principle only in those instances where it is expressly mentioned; 2—is not a rule opposed to the civil reckoning, to be applied only in acquisitions or in the duration of the *status facti;* 3—but is a benevolent general interpretation of the civil reckoning, and found as such in the sources (Cf. D. [50, 16] 134); 4—is not therefore applicable only in those cases in which it would inflict damage to the person on whom the term is imposed, as in the case of the *decadentia,* and of the prescriptions which are extinctive of rights.[20]

According to this same author the application of the rule *postremus totus* is found in the following cases:[21]

A—In the prescription of temporal actions;[22]

B—In the *bonorum possessio.*[23]

The rule *dies coeptus* is applied in the following instances:

A—In the computation of the year for the *anniculus;*[24]

[19] This was the interpretation commonly accepted by canonists. Cf. Ojetti, *Synopsis,* s.v. *Tempus;* Maroto, *Institutiones Iuris Canonici* ad *Normam Novi Codicis,* Vol. I, (3. ed., Romae: Apud "Commentarium pro Religiosis," 1921), n. 257, 5. This work will hereafter be cited as *Institutiones.* Vinnius writes: "Glossa notat diem coeptum pro completo haberi, quoties id alicui utile est, aut alioqui favorabile, ut in usucapionibus, 1. 6 *de usucap.* in testamenti factione, leg. 5. *qui testam. fac.* in manumissionibus, leg. 1. *de manumiss* quo etiam pertinet *leg. anniculus* 134. *de verb. signif.* In honoribus quoque favoris gratia constitutum esse, ut annus inchoatus pro pleno habeatur, *leg.* 8 *de mun. et hon.* Contra autem, ubi favorabile est, non numero dierum, sed momentis temporum annum dividi; diem coeptum pro completo non haberi, ut in temporalium actionum praescriptione, *leg.* 6. *de oblig. et act., in vacatione tutelae, leg.* 2. *de excus."—Institutionum Imperialium Commentarius,* lib. I, tit. XXIII, n. 4.

[20] *Ius Digestorum,* I, n. 841, I, d.

[21] Cf. *op. cit.,* I, n. 841, II, a.

[22] "In omnibus temporalibus actionibus, nisi novissimus totus dies compleatur, non finit obligationem." D. (44, 7) 6.

[23] Cf. D. (38, 9) 1, 9.

[24] "Anniculus non statim ut natus est, sed trecentesimo sexagesimo

B—In the reckoning of the age required to make a will;[25]
C—For the capacity of manumitting.[26]

§5—The *Dies Intercalaris*

Another knotty problem was that of the *dies intercalaris.* The Roman calendar did not consider the *bissextus Kalendas Martii* as a separate day, but simply looked upon the *sextus* and the *bissextus* as one and the same day.[27]

Now, authors are not of one mind concerning the day actually intercalated. Some are of the opinion that it corresponded to our February 24th, while others say it was February 25th.[28]

From the above mentioned text February 24th appears to be the intercalated day since it corresponds to the *posterior dies.* That is clearly shown by the Roman system of counting the days of the month.

Thus Feb. 29 corresponded to *Pridie Kal. Martii*

28..............*Tertius ante Kal Martii*
27..............*Quartus*..............
26..............*Quintus*..............
25..............*Sextus prior*..........
24..............*Sextus posterior*-(*Dies intercalaris*)
23..............*Septimus* ....etc.[29]

quinto die dicitur, incipiente plane, non exacto die, quia annum civiliter non ad momenta temporum sed ad dies numeramus."—D. (50, 16) 134.

[25] Cf. D. (28, 1) 5.

[26] Cf. D. (40, 1) 1.—This rule possibly applies also to *usucapio.* The texts, however, are obscure and interpolated. For a more complete exposition of this problem cf. D'Angelo, *op cit.,* I, n. 841, II, c.. d, e; Windscheid, *Pandette,* I, #103, 2, a, —2.

[27] "Proinde et si bissexto natus est, sive priore sive posteriore die, Celsus scripsit nihil referre: nam id biduum pro uno die habetur et posterior dies Kalendarum intercalatur."—D. (4, 4) 3 #3. Cf. D. (50, 16) 98; D. (44, 3) 2. In other words the *sextus Kalendas Martii* was simply a day of 48 hours. Cf. Wernz-Vidal, *Normae Generales,* n. 242, IV, note 6.

[28] Cf. Windscheid, *Pandette,* I, #103, 2, b; D'Angelo, *Ius Digestorum,* I, n. 835; Ojetti, *Normae Generales,* p. 195, notes 8-9.

[29] Cf. Maynz, *Cours de Droit Romain,* I, #125, 3.

From this one sees that the legal length of the leap year with the Romans was 365 days only, and that February always had 28 days, never 29. This technical reckoning, however, applied only to the time-periods prescribed by the law. Individuals could consider this *dies bissextus* as a distinct day in contracts if they so desired.[30]

§6—Special Reckoning of the Year and the Month

A final remark must here be made concerning these two time-units. In a few cases the year was counted as complete the moment it began.[31] The same rule also occasionally applied to the month.[32]

Article 3—*Tempus Utile*

Another Roman institute one has to reckon with considered time either as *utile* or *continuum*. Time was known as *utile* when it did not lapse if obstacles or hindrances of an excusing nature prevented one from acting. Obstacles were divided into two general categories: lack of knowledge of one's rights, and impossibility of acting.[33]

[30] Cf. Windscheid, *Pandette,* I, #103, 2, b.

[31] "Annus autem vicesimus quintus pro completo habetur: hoc enim in honoribus favoris causa constitutum est, ut pro plenis inchoatos accipiamus."—D. (50, 4) 8. Cf. D. (36, 1) 74 #1.

[32] "De eo textu autem, qui centesimo octogesimo secundo die natus est, Hippocrates scripsit et divus Pius pontificibus rescripsit iusto tempore videri natum, nec videri in servitutem conceptum, cum mater ipsius ante centesimum octogesimum secundum diem esset manumissa."—D. (38, 16) 3 #12.—"Septimo mense nasci perfectum partum iam receptum est propter auctoritatem doctissimi viri Hippocratis: et ideo credendum est eum, qui ex iustis nuptiis septimo mense natus est, iustum filium esse."—D. (1, 5) 12.

[33] "Ita autem utile tempus est, ut singuli dies in eo utiles sint, scilicet, ut per singulos dies et scierit et potuerit admittere (bonorum possessionem): ceterum, quacumque die nescierit aut non potuerit, nulla dubitatio est, quin dies ei non cedat."—D. (38, 15) 2. "Scientiam autem eam observandam Pomponius ait, non quae cadit in iuris prudentes, sed quam quis aut per se aut per alios adsequi potuit, scilicet consulendo prudentiores, ut dilegentiorem patrem familias consulere dignum sit."—*Loc. cit.*

Thus the days that were granted for appeals did not count if the person in question was prevented from so doing.[34]

Tempus continuum, on the other hand, was not affected by the hindrances that might have prevented one from using one's rights. The days, months and years were counted in an obsolute way as nature meted them out. In the *usucapio,* for instance, time was always reckoned as *continuum.*[35]

The notion of *tempus utile* or available time has slightly changed with the years in Roman Law. In later periods time was said to be *utile* if it did not lapse when one could not actually use it. The same concept exists today. If ten days of *tempus utile,* for instance, are given for an appeal, and the person in question is hindered from so doing during two days, these two days are not counted and two extra days are added, but this is done only once it becomes clear that a legal impediment really prevented the person from appealing. Originally, on the contrary, *dies utiles* were granted independently of one's subjective ability of making use of the allotted time. Court holidays seem to have been the only accepted impediments. Thus, if one was given ten days to appeal, and it was foreseen that no sessions would be held during five of these days, the praetor immediately granted an extension of five days which could also be *used* to appeal. Thus one sees that the denomination of *utile* was applied, originally, only to the extra time granted.[36]

[34] "Quia tractatus de utilibus diebus frequens est, videamus quid sit experiundi potestatem habere. Et quidem in primis exigendum est, ut sit facultas agendi . . . Proinde sive apud hostes sit, sive rei publicae causa absit, sive in vinculis sit aut si tempestate in loco aliquo vel in regione detineatur, ut neque experiundi potestatem non habet. . . Illud utique neminem fugit, experiundi potestatem non habere eum, qui praetoris copiam non habuit."—D. (44, 3) 1. Cf. D. (48, 50) 12 [11] #5.

[35] "In usucapionibus mobilium continuum tempus numeratur."—D. (41, 3) 31 #1.

[36] Maynz (*Cours de Droit Romain,* I, #126) thus explains the origin of the expression *tempus utile*: "Dans l'ancien droit, le nombre des jours auxquels it était permis d'agir en justice était fort restreint. Il en résultait souvent qu'un espace de temps assez long n'offrait en réalité aux parties qu'un petit nombre de jours, puisque de fait tous les jours néfastes étaient inutiles dans le délai. Cela était surtout sensible dans les

Tempus continuum was always the rule; *tempus utile*, considered as exceptional, had to be expressly granted.[37]

Tempus utile was conceded only for impediments or obstacles of a short duration. Other provisions were made for such cases as minority, insanity, etc.[38]

Among the obstacles that prevented *tempus utile* from lapsing, the impossibility of reaching the judge was the most common. Mention is also made of incarceration, war, absence for the service of the State, detention, illness, physical impossibility of reaching one's adversary. Sundays and court holidays were most likely not considered legitimate impediments.[39]

délais, ordinairement tres courts, que les préteurs fixaient pour l'exercice des actions nouvelles ou des autres remèdes de droit qu'ils créaient. Pour ne pas contrarier l'équité, dont ils se rendaient ordinairement les interprètes, ces magistrats décomptaient donc les jours inutiles, et ne faisaient entrer dans les délais que les jours où les parties étaient réellement à même de s'adresser à la justice. Les délais qu'on comptait ainsi étaient appelés *utiles*, tandis qu'on donnait le nom de *continuum tempus* à un délai dans lequel tous les jours sans distinction étaient comptés. Par la suite, quand la distinction dont il vient d'être parlé eut disparu, on nomma *tempus utile* celui dans lequel on ne comptait que les jours auxquels la partie avait pu agir en justice, *experiundi potestatem habere*, en négligeant les jours où, par un obstacle quelconque, elle n'avait pas joui de cette faculté."

[37] Cf. Maynz, *loc. cit.; * Wernz-Vidal, *Normae Generales*, n. 250. Donellus writes: "Annum autem, et diem cum simpliciter dicunt Iurisconsulti, utilem accipi vult Govean. I. *Varior. Lect.* 10. Sed contra vulgo, et usu forensi recepta est sententia, tempus in dubio accipi de continuo, nisi expresse dicatur, quod sit utile. Praeter alios Coler. de *Process. exsec.* Par. 4. c. I, num. 200."—*Opera Omnia Commentariorum de Iure Civili* (12 tomes, Romae, 1828), *De Iure Civili*, I, col. 145, note 1.

Concerning the instances when *tempus utile* was granted the reader is referred to D'Angelo, *Ius Digestorum*, I, n. 843, a, b; Ojetti, *Normae Generales*, p. 205, note 4. Van Hove (*De Temporis Supputatione*, n. 317) writes: "Praetor tribuebat tempus utile, ad exercendas actiones temporarias quas concedebat, vel ad petendam bonorum possessionem. Tempus utile etiam intelligebatur concessum heredi qui per cretionem, seu declarationem sollemnem, acceptare debebat hereditatem, vel ad petendam restitutionem in integrum."

[38] Cf. Windscheid, *Pandette*, I, #104, note 5.

[39] Cf. Windscheid, *op. cit.*, I, #104, note 6; D'Angelo, *Ius Digestorum*, I, n. 843, c. Cf. also D. (44, 3) 1.

The question of ignorance with reference to *tempus utile* has given birth to many controversies. These have, above all, centered around the admissibility in Roman Law of a certain combination of *tempus utile* and *tempus continuum* called *tempus utile ratione initii et continuum ratione cursus*. According to this form of reckoning, the only impediment acknowledged by law to prevent one's allotted time from lapsing was the state of ignorance concerning the possession of one's rights. The moment that ignorance disappeared the time automatically became *continuum*. Thus the ten days given to appeal would not lapse before the official notification that an adverse sentence had been rendered.[40]

Mention is also made of the *tempus continuum ratione initii*. This referred to a certain amount of time that began to lapse on a determinate day without due consideration of the fact that ignorance intervened or not.[41]

The reckoning of the obstacles and hindrances relative to *tempus utile* must necessarily be made in a moral way, which gives rise to many an incertitude. Justinian (483-565 A.D.) therefore strove to limit this *tempus utile*, and one learns that

[40] Cf. Maynz, *Cours de Droit Romain*, I, #126; Wernz-Vidal, *Normae Generales*, n. 250.—Others deny the existence of this form of time-reckoning. According to them time is either *continuum* or *utile*, and nothing else. In the cases here referred to the time is *continuum*, but the starting point of the computation is determined by the knowledge one has of one's capacity of acting: *tempus decurrit a momento scientiae*.—Cf. D'Angelo, *Ius Digestorum*, I, n. 842.

[41] Cf. Maynz, *Cours de Droit Romain*, I, #126. This same writer (*loc. cit.*, note 3) states that a thorough analysis of the various possibilities reveals the following forms of time reckoning: *tempus continuum ratione initii et ratione cursus*, the ordinary time-interval; *tempus continuum ratione initii, utile ratione cursus; tempus utile ratione initii, continuum ratione cursus; tempus utile ratione initii et ratione cursus*.—Cf. Wernz-Vidal, *Normae Generales*, n. 250.—For a thorough examination of the controversy the reader is referred to Windscheid, *Pandette*, I, #104, note 7.

he changed the *tempus utile* of one year (*annus utilis*) for the *restitutio in integrum*, to four years *tempus continuum*.[42]

[42] "Supervacuam differentiam utilis anni in integrum restitutionis a nostra re publica separantes sancimus et in antiqua Roma et in hac alma urbe et in Italia et in aliis provinciis quadriennium continuum tantummodo numerari ex die, ex quo annus utilis currebat, et id tempus totius loci esse commune: ex differentia enim locorum aliquod induci discrimen satis nobis absurdum esse visum est."—C. (2, 52 [53]) 7.—Cf. Maynz. *Cours de Droit Romain*, I, #126, note 1. This author adds that it would have been more rational to extend this change to all juridical institutes instead of restricting it to the *restitutio in integrum.*

Chapter V

CANON LAW

Article 1—Preliminary Remarks

The reckoning of time in Canon Law proceeds directly from the Roman Law.[1] Before the Code the authors invariably substantiated their statements in this matter by referring to some text of Roman Law.[2] That is easily understood if one considers for an instant that the Church in adopting the Roman calendar was, so to say, forced to accept the Roman standards of time-reckoning. These are two inseparables; one cannot be thought of without the other, no more than it is possible to imagine a gun firing shells of a different caliber.

Such has been the common teaching of the doctors.[3] However, special rules have occasionally been prescribed by the Church, but these have been very rare and refer, almost without exception, to the reckoning of the hours of the day. The great majority of these special laws were furthermore given towards the end of the last century.[4] In other matters of a purely disciplinary character the only clear intervention of the Church was given when the publication of the Code was well under way. It pre-

[1] Cf. *supra,* p. 51, for the few exceptions which are of Jewish origin.

[2] Cf. D'Annibale, *Summula Theologiae Moralis* (2. ed., 3 vols., Mediolani, 1881-1883), I, n. 39. This work will hereafter be referred to as *Summula.*

[3] Van Hove (*De Temporis Supputatione,* n. 272) writes: "In iure canonico, praeterquam in liturgia et in Martyrologio in quibus Ecclesia usurpabat et adhuc hodie adhibet annum lunarem, prius admissum est *calendarium* Iulianum, deinde calendarium Gregorianum seu Iulianum a Gregorio XIII anno 1582 reformatum. *Principia* Ecclesia mutuata est a iure Romano, at cum diversa in diversis textibus legerentur, de eius interpretatione et applicatione doctrinae valde diversae propositae sunt. Insuper, in nonnullis materiis, ipsa speciales regulas supputationis determinavit. Inde mira opinionum diversitas, multae obscuritates et dubia plurima in iure canonico orta sunt."—Cf. Wernz-Vidal, *Normae Generales,* n. 241.

[4] Cf. *infra,* pp. 94-100.

scribed the so called civil reckoning for the year of novitiate.[5] That decision, at least of its own intrinsic value, applied only to that institute, and not to similar cases.

In other words there were, prior to the promulgation of the Code, no general norms for the reckoning of time, such as one now finds in canons 31-35, and, logically enough, the matter was not treated under one heading by the authors. The question was considered here and there, in those treatises where the element of time was involved, as in the title *de feriis,* for instance.[6] Logically enough variations were found in the divers institutes, and the commentators were not agreed upon the course to follow even when a single question was considered.[7]

One principle, however, seems to have guided the authors in the reckoning of time, and that was to apply advantageously as often as possible the famous rule: *"Odia restringi, et favores convenit ampliari."*[8] That should clearly appear from the following pages where the distinction between favorable and odious matters is more than once invoked.[9]

Another obstacle that one has to hurdle in the proper understanding of the historical aspect of time-reckoning in the Church is the loose terminology occasionally found. The word *dies,* for instance, needs careful consideration. According to the Romans the natural day, *dies naturalis,* was defined as the interval of time between dawn and dusk, and was opposed to night, *nox.*[10] Such is the meaning that Gonzalez-Tellez attributes to the same word. According to him the *dies civilis* is reckoned from midnight to

[5] S.C. de Religiosis, decr. 2 maii 1914, n. 1.—*Fontes,* n. 4420.

[6] Cf. Wernz-Vidal, *Normae Generales,* n. 241. A few authors, it is true, had written separate works on time, such as Antonellus, *Tractatus Novissimus et Absolutissimus de Tempore Legali,* but casuistry made up the greater part of the contents. There was no synthetic presentation of the matter, perhaps because there were no general rules as they now obtain in Title III of Book I of the Code. For more details concerning authors that wrote on legal time Cf. Ojetti, *Normae Generales,* p. 191.

[7] Cf. Van Hove, *De Temporis Supputatione,* n. 274.

[8] Reg. XV., R. J. in VI°.

[9] Cf. *infra,* pp. 84-87.

[10] Cf. *supra,* pp. 38-39.

midnight.[11] Antonellus, on the other hand, states the exact opposite. He maintains that the natural day is composed of 24 hours reckoned from midnight to midnight,[12] but that the civil day is reckoned from dawn to dusk.[13]

The words *ortus solis* have also been given a rather elastic signification. Schmalzgrueber (†1735) speaks of a month beginning with the rising sun—*ortus solis*—which, however, happens to be just another name for midnight.[14]

The expression *tempus verum* has also had diverse meanings.[15]

Article 2—The Duration of the Year and the Month

§1—The Year

This time-unit has always consisted of 365 days when considered in general.[16] The question of the extra day of the leap year, which cannot even now be looked upon as completely settled, has proved to be a delicate problem throughout the ages. There is on this point an official pronouncement of the Church, at least as far as liturgy is concerned. Pope Alexander III (1159-1181) declared that the 24th and the 25th of February were practically one an the same day.[17]

[11] *Commentaria Perpetua in Singulos Textus quinque Librorum Decretalium Gregorii IX* (4 vols., Venetiis, 1699), lib. I, tit. XXIX, c. XXIV, n. 3. Hereafter this work is cited as *Commentaria Perpetua.*

[12] *De Tempore Legali,* lib. I, c. III, n. 27.

[13] *Op. cit. lib.* I, c. III, n. 31. Cf. Calà, *Tractatus Absolutissimus de Feriis, Solennibus, Repentinis, et Indictis* (Neapoli, 1675), Quaestio IV, nn. 308, 321, 322. This work will hereafter be referred to as *De Feriis.*

[14] *Ius Ecclesiasticum Universum* (12 vols., Romae, 1843-1845), lib. III, tit. V, n. 305. Cf. Suarez, *Opera Omnia* (28 tomes, Parisiis, 1856-1878), *De Diebus Festis,* lib. II, c. XXXI, n. 6.

[15] Cf. *infra,* pp. 97-98.

[16] Cf. Van Hove, *De Temporis Supputatione,* n. 273; Antonellus, *De Tempore Legali,* lib. I, c. III, nn. 7-8; Tiraquellus, *Commentaria de Utroque Retractu, et Municipali, et Conventionali* (4. ed., Venetiis, 1562), *De Retraict. Lignagier.* #1, *Gloss.* X, nn. 2-3. This work will hereafter be cited as *De Utroque Retractu.*

[17] "Festum vero beati Matthiae iuxta consuetudinem ecclesiasticam vigilia eatenus praecedat, ut nec pro bissexto, nec pro quolibet alio modo inter

But even this statement did not put an end to the controversy, since the two days involved are not considered by him exactly one and the same day as was the case with the Roman Law. Here the expression is modified *"qui duo quasi pro uno reputantur."* The feast of St. Matthias is to be celebrated either on the 24th or on the 25th of February, that is, during 24 hours only and not 48, *disjunctive* and not *conjunctive.*[18]

There seems to be one general rule, however, according to which the *sextus* and the *bissextus* Kalendas Martii were always considered as two distinct days, and computed as such. That happened when a period of time was expressed in days and not in years or months.[19]

Similarly when February and the leap year were explicitly mentioned as such, these units were reckoned as they actually occurred in the calendar. Thus, if any one were suspended for the month of February, and the year happened to be a leap year, the suspension would last 29 days. But if one were suspended for a month on February 1 without explicit mention of that month, then one could celebrate on February 29th.[20]

The second general trend that one notices in the writings of the authors up to the end of the last century is that they distinguish between favorable and odious matters. The more common opinion

se et solennitatem aliam diem admittat, in qua utique vigilia, nisi venerit in dominica die, ieiunium celebretur. Ipsum autem festum sive fiat in praecedenti die sive in sequenti, qui duo quasi pro uno reputantur, nullus error, sed consuetudo ecclesiae teneatur."—C. 14, X, *de verborum significatione,* V, 40.

[18] Cf. Suarez, *De Censuris,* disp. XXIX, sectio I, n. 8. The same interpretation was also applied to other canonical institutes. Thus the obligation of fasting on the last day of February would fall either on the 28th or, in leap years, on the 29th. Cf. *loc. cit.*

[19] "Hoc tamen intellige, quando computantur menses, vel anni: nam si computentur dies, ut si dicatur, Sis suspensus 40. diebus, computabitur dies bissextus: quia est verus dies: et sic erunt illi duo dies . . . "—Sanchez, *De Sancto Matrimonii Sacramento* (3 tomes, Antverpiae, 1626), lib. II, disp. XXIV, n. 18. This work will hereafter be referred to as *De Matrimonio.* Cf. Calà, *De Feriis,* q. IV, n. 691.

[20] Cf. Sanchez, *loc. cit.*

seems to have maintained that the two days were considered as one in favorable affairs, but as two distinct days in odious matters. Thus a vacation given for three months from 10 A.M. November 24th would not have expired before the recurrence of the same hour February 25th of the leap years. A suspension, however, imposed at the same time on some one else, would have come to an end at 10 A.M. February 24th.[21]

In the reckoning of the age required for ordinations the leap year day could not be excluded. Such, at least, was the more common teaching of the authors.[22]

However, this Roman Law fiction lost favor with the canonists as time sped along. The authors adhering to it at the end of the last century could easily be counted. D'Annibale, for instance, simply does away completely with this inconvenience. He states

[21] Cf. Sanchez, *De Matrimonio,* lib. II, disp. XXIV, nn. 17-18. Cf. also Suarez, *De Censuris,* disp. XXIX, sectio I, n. 7, where it is said that in odious things a year is always 365 days, and never 366, not even in leap years; Tiraquellus, *De Utroque Retractu, De Retraict. Lignagier.* #1, *Gloss.* X, n. 123.

In Calà (*De Feriis,* q. IV, n. 663) a slightly different view is expressed.

[22] The reception of orders is a favor, yet it involves the assumption of obligations which can not be classed among the *favorabilia.* Hence, the possibility of assuming these obligations should be postponed rather than anticipated, which effect can be gained only when the *dies bissextus* is counted along in the years past which one must be in order to receive orders. Cf. Ferraris: "In supputanda ordinandorum aetate dies anni bissextilis non potest excludi, iuxta communiorem et veriorem sententiam: Bon. d. 8. q. unic., p. 5, n. 14; Sanch. de Matr. d. 24, n. 22. Escob. Diana, aliique plures cum D. Alphons."—*Bibliotheca Canonica Juridica Moralis Theologica* (ed. novissima, 9 tomes, Romae: 1885-1899), s.v. *Aetas,* n. 53. This work will hereafter be cited as *Bibliotheca.*

Cf. also St. Alphonsus, *Theologia Moralis* (9 vols., Venuntiae, 1828), lib. VI, n. 800, for other opinions.—Engel (*Collegium Universi Juris Canonici* [9. ed., Beneventi, 1760], lib. I, tit. XIV, n. 5) and Barbosa (cited by Engel, *loc. cit.*), for instance, hold that this leap year day is not counted. One could infer that these authors do not accept the distinction between favorable and odious matters, but that they strictly adhere to the old Roman system.

that the leap year always consists of 366 days, since the month of February is no longer to be considered as intercalated.[23]

On the other hand Vermeersch did not abandon this old Roman relic till a few years after the Code. He seems to have adhered to the more rigid view that the Roman system was to be followed in odious matters as well as in favorable ones.[24]

§2—The Month

The abstract, theoretical value of this time-unit never changed. It always was 30 days. In practice, however, every imaginable reckoning has at some time or other found favor with the authors. Antonellus tells us that, according to some writers, the legal month is invariably to consist of 30 days. Others, on the other hand, maintain that the months are to be computed as they occur in the calendar (*computatio prout currunt*). Thus one month from Jan. 10 would run to Feb. 10, and one month from Feb. 15 would come to an end March 15, etc. Another school combines these two theories and contends that the calendar reckoning is to be followed if there is a determined starting point for the time-interval to be computed. It matters little whether the starting point is determined *a iure* or *ab homine*. Conversely, when there is no such determination, the length of any of the twelve months is considered to be 30 days.[25]

Antonellus has an opinion of his own. When *tempus utile* is involved, the month always has 30 days. However, if a time-interval is to be reckoned as *tempus continuum*, the calendar computation is to be adhered to, January having 31 days, February 28 days, etc.[26]

[23] "Et bissextus 366; non est enim Februarius amplius mensis intercalaris."—*Summula*, I, n. 39. Cf. Ojetti, *Synopsis*, s.v. *Tempus*.

[24] Cf. Vermeersch, "De Ratione Anni Bissextilis Habendae in Recta Temporis Computatione,"—*Periodica de Re Canonica et Morali utili praesertim Religiosis et Missionariis*, I (1911), 3-5. This article treats especially of the reckoning of the year prescribed for the novitiate. Vermeersch refers to many authors, e.g. Lessius, Molina, Sanchez, Barbosa.—The periodical here referred to will hereafter be cited by the word *Periodica*.

[25] *De Tempore Legali*, lib. I, c. III, nn. 15-17.

[26] *Op. cit.*, lib. I, c. III, nn. 18-19.

Sanchez states unhesitatingly that the calendar reckoning is to be followed when there is a determinate starting point.[27] And such seems to have been the common teaching in later years.[28]

But multiple theories appeared concerning the case wherein the starting point was not determined. Sanchez stated that, according to some authors, the month was then to consist of 28 days, but that 30 was the number chosen by other canonists. According to another theory the matter was to be left to the judgment of a prudent person (*prudentis arbitrio relinqui*).[29]

Lugo, for instance, stated that even then the month was to be reckoned according to the calendar.[30] Finally, others maintained that the month then meant exactly the twelfth part of the year.[31]

In later years at least one finds the famous distinction between favorable and odious matters also in reckoning months. It was then admitted that in odious things a month should not have more than 30 days.[32]

The reckoning of the month according to the calendar brings about the question: which calendar is then to be followed? The present writer does not know of any official ecclesiastical document concerning the calendar before the end of the 19th century, if one excepts the question of the leap year day. The matter was however considered by a few authors. The common opinion was that the faithful had to follow the Gregorian calendar in those regions where it was in use. Elsewhere, one was free to adopt the Julian calendar.[33]

[27] Cf. *De Matrimonio,* lib. II, disp. XXIV, n. 15.

[28] Cf. Ojetti, *Synopsis,* s.v. *Tempus;* D'Annibale, *Summula,* I, n. 39. Cf. also Calà, *De Feriis,* q. IV, n. 571.

[29] *De Matrimonio,* lib. II, disp. XXIV, n. 14.

[30] *Disputationes Scholasticae et Morales* (ed. nova, 8 vols., Parisiis, 1868-1869)—*De Justitia et Jure,* disp. XXII, sect. XII, n. 314.

[31] Thus one reads in Calà (*De Feriis,* q. IV, n. 572) : "Si vero jus non exprimat, a quo tempore incipiat mensis, et tunc assumitur duodecima pars anni."

[32] Cf. Van Hove, *De Temporis Supputatione,* n. 273, where reference is made to D'Annibale, *Summula,* I (5. ed.), n. 39, note 9.

[33] La Croix († 1714) wrote: "Ubi servatur vetus calendarium, Ecclesia permittit, ut fideles se illi accommodent, uti constat ex usu piorum et

The question of the lunar calendar was not even mentioned before 1890. It was then asked of the Holy Office whether or not the years in the age required for the contracting of a valid marriage were necessarily solar years, the lunar years being excluded. The answer stipulated that those years were solar years. When it was impossible to reckon one's age in terms of solar years, an extra month was to be added to the lunar years. However, this provision was not to affect those marriages that had already been contracted.[34]

Theoretically, two other units, the week and the day, might be mentioned at this point. The week, however, was practically nonexistent as a juridical entity. Its reckoning was furthermore without difficulty and thus deserves no specific consideration. The day presents a problem of its own and is treated in Article 5.[35]

Article 3—The Reckoning of Time

The various forms of time-reckoning used by the Romans were adopted by the Church. One finds the natural and the civil reckoning just as in the Roman Law, and not in any way better defined.[36]

A logical examination of the question demands that one inquire, first of all, when the natural reckoning was to be used, and when the civil was to be applied. In the event that the latter applies one must then delve into the problem of the *dies terminus a quo* and

doctorum hominum; ubi autem receptum est novum, secundum illud est procedendum: et hic valet illud."—*Theologia Moralis* (2 tomes, Coloniae, 1739), lib. I, n. 585.

[34] "Si responda al proponente che quanto al passato taccia; per l'avvenire poi ritenga la massima che gli anni richiesti dai sacri canoni per la validità del matrimonio sono anni solari, non lunari, e da computarsi da giorno a giorno in cui si celebra il matrimonio. Che si ciò non possa farsi, si computi l'età secondo l'anno lunare aggiundendovi un mese."—S.C.S. Off. (Chan-si), 7 maii 1890—*Fontes*, n. 1122. The Code has not touched upon this topic, which therefore remains as it was in 1890. The question, actual as it is, will advantageously be discussed in the commentary. Cf. *infra*, pp. 195-199.

[35] Cf. *infra*, pp. 89-101.

[36] Cf. Van Hove, *De Temporis Supputatione*, n. 274.

the *dies terminus ad quem*. In the reckoning for instance, of the year prescribed for the novitiate, was the day of the vesting, incomplete though it was, considered as one full juridical day? Was one said to have reached the age of fourteen with the beginning of one's fourteenth birthday anniversary, or only at the end of that day? [37]

§1—The Natural Reckoning [38]

Various have been the opinions of the authors concerning the use of the moment to moment computation of time in Canon Law. According to many eminent writers the natural reckoning was the rule. This is at least implicitly stated by Gonzalez-Tellez.[39] According to Ioannes Andreae (†1348) and Panormitanus (†1453) the natural computation was to be used unless the civil reckoning was expressly provided for by law.[40]

Other authors were less emphatic. The natural reckoning was to be followed in certain instances only. According to Reiffenstuel (†1703) two general norms governed the use of this form of computation. In the first place the natural reckoning was to be followed when the prolongation of a time-interval was looked upon as favorable. Otherwise one had to follow the rule *Dies coepta habetur pro completa*.[41] The second norm stipulated that the natural reckoning was to be used when an interval of time was to be computed from some determinate act.[42]

[37] The problem is not always thus considered by the authors. Very often the natural reckoning is not squarely opposed to the civil computation as such.

[38] Cf. Wernz-Vidal, *Normae Generales*, n. 244.

[39] Cf. *Commentaria Perpetua*, lib. I, tit. XXIX, c. XXIV, n. 3.

[40] Cf. Reiffenstuel, *Ius Canonicum Universum* (4 tomes, Venetiis, 1735), lib. II, tit. XXVII, n. 111.

[41] *Op. cit.*, lib. III, tit. XXXI, n. 95. Other canonists held a different view and considered the problem from another angle. Cf. Calà (*De Feriis*, q. IV, n. 696): "Verum Alciat. distinguit sic: nam aut tempus sumitur in his, quae favorem respiciunt, et in his satis est diem esse coeptum. . . .aut sumus in odiosis, et de momento ad momentum fit computatio."

[42] *Loc. cit.*—This was substantially the opinion held by Passerinus. He stated that the natural reckoning was to be used when a time-interval was

Sanchez considered the matter from another angle. He held that the natural reckoning was regularly used when the starting point was a determinate day.[43]

According to others the natural reckoning was used in penal affairs, while the civil was applied in judicial matters.[44]

Lega (†1935) held a view directly opposed to that of Ioannes Andreae and Panormitanus. He maintained that the natural reckoning was exceptional, and was to be used only when it was prescribed by positive law, as was the case of the ten days given for one to appeal. These ten days began to lapse from the moment the sentence was given and not from the following day. But in the course of time canonical equity modified this. An appeal could then be made up to the last moment of the tenth calendrical day after that on which the sentence had been given.[45]

§2—The Problem of the Last Year or Month

When the civil reckoning was used the fundamental problem was to decide whether or not the last year or month mentioned was to be actually completed. It must here be recalled that, when honors were granted, the Roman Law considered the last year of one's age as complete as soon as it had begun.[46] Such a form of reckoning was also used by the Church but not precisely for the same reason. In the Roman Law the nature of the institute involved was the prime factor in the use of this form of reckoning. Conversely, in ecclesiastical affairs the wording of the law itself, especially in later years, determined whether or not the last

computed, not from a certain day, but from a certain act (*ab actu aliquo expleto*). Cf. Wernz-Vidal, *Normae Generales* n. 244, note 12.

[43] *De Matrimonio,* lib. II, disp. XXIV, n. 16. Cf. also Tiraquellus, *De Utroque Retractu, De Retraict. Lignagier,* #1, *Gloss.* XI, n. 17.

[44] Cf. Calà, *De Feriis,* q. IV, nn. 491 and 615.

[45] Cf. *Praelectiones in Textum Iuris Canonici de Iudiciis Ecclesiasticis* (4 vols., Romae, 1896-1901), I, 405.—This work will hereafter be referred to as *De Iudiciis.* Cf. Wernz-Vidal, *Normae Generales,* n. 244, note 12—De Angelis is of the same opinion as Lega. Cf. *Praelectiones Iuris Canonici ad Methodum Decretalium Gregorii IX Exactae* (4 vols., Romae, 1877-1887), lib. IV, tit. VIII, n. 9.

[46] Cf. *supra,* p. 66.

year or month was to be actually completed. However, it must be said that many divergent opinions were propounded relative to this question.

One reads in the Decree of Gratian that the subdiaconate was not to be conferred before the age of twenty.[47] The text itself suggested that the twentieth year was to be completed. Nevertheless the Glossa does not consider this, but explains the fact that the twentieth year must be completed because the rule according to which a year begun is held as complete applied to civil and not to ecclesiastical honors, or perhaps because obligations were attached to ecclesiastical honors.[48]

In the Decretals slightly different explanations are found. It was stated that no one was to be chosen as bishop before the completion of his thirtieth year.[49] Here again the Glossa distinguished between civil and ecclesiastical honors, but also added another consideration. It stated that the ruling for the choice of bishops was possibly exceptional and, consequently, that in the reception of the inferior dignities it was sufficient for one to have *reached* (*attingere*) the age prescribed by law.[50]

One can easily conclude from the above mentioned norm relative to bishops, that the wording itself cannot be interpreted in any other way than to mean that the thirtieth year must be actually completed. From then on the predominant consideration in this matter seems to be the text of the law and no longer its nature.

Antonellus left us a fair account of the general teaching of his time concerning the matter under examination. It was commonly admitted that a year in law meant a full year. There were, however, many exceptions to this rule.

Two of these special provisions were pure Roman Law. The first of these stipulated that (civil) honors and charges could be

[47] "Subdiaconus non minor viginti annorum ordinetur."—C. 4, D. 77.

[48] *Loc. cit.*

[49] C. 7, X, *de electione et electi potestate*, I, 6.

[50] *Loc. cit.* The same considerations are also found in the *Glossa* (c. 8, X, *de regularibus et transeuntibus ad religionem*, III, 31), s.v. *ante consummationem*.

granted to those who had reached their 25th year. The second stated that it was sufficient to be in one's 21st year in matters favorable to heirs.

In purely ecclesiastical affairs the prescribed age for the obtaining of simple benefices, and of canonicates in collegiate churches, was the beginning of one's 14th year. Benefices to which was attached the care of souls were granted only to those candidates who had reached their 25th year.

In other instances the legal text was considered. Thus, if the law prescribed that a certain thing could be performed in one's 18th or 20th year, it was then held sufficient to attain to one's 18th or 20th year. Such expressions as *iuxta, in, ad, intra, ante,* meant that the year mentioned did not have to be a complete or full year. The same conclusion was held when the law used the ablative case. Conversely, full years were required when the law used the genitive case, or the accusative case with the prepositions *per* or *post.* The term *usque* was of a more difficult interpretation. It was generally admitted that in favorable affairs the computation of a full year was not required. On the other hand the full amount of time had to elapse in odious matters. However, the Rota has been known to render two contrary decisions relative to the interpretation of the word *usque.* Antonellus' own opinion was to leave the decision to the judge.

Finally a few authors maintained as a general principle that the rule *annus coeptus habetur pro completo* was to be followed in all favorable affairs.[51]

[51] Cf. Antonellus, *De Tempore Legali,* lib. IV, c. I, nn. 7-19. Cf. also Panormitanus, *Opera Omnia, super Primo Clement. De aeta. et equal.* Cap. II, 6; Decius, *In Decretales Commentaria* (Lugduni, 1576), *cum in cunctis, De elect.,* n. 48.—For concrete applications of this rule in more recent years cf. Ferraris, *Biblotheca,* s.v. *Aetas,* nn. 4, 10, 17, 23, where he mentions the subdiaconate and the simple benefices. The age required for pastors was another example. Cf. Wernz, *Ius Decretalium* (2 ed., 6 vols., Romae, 1905-1913), II, n. 112, IV—For the reception of minor orders the years required always had to be actually completed. Cf. Ferraris, *op. cit.,* s.v. *Aetas,* n. 3; Pirhing, *Ius Canonicum in V Libros Decretalium* (4 vols., Dilingae, 1722), lib. I, tit. XI, n. 11.—The 40 years required for the abbess were also complete years. Cf. Ferraris, *op. cit.* s.v. *Abbatissa.*

It must be borne in mind that this form of reckoning the year never considers a day which is begun as one that is already completed. Thus the beginning of the 20th year demands that the last day of the 19th year be completed. The reason is that the beginning of the last day of the 19th year would mark the legal end of that year, at least in favorable matters, but it would not mark the beginning of the 20th year which is here alone considered. Thus a subdeacon who would have reached the prescribed age for the diaconate at four P.M. December 21st (natural reckoning), could not have been ordained during the morning of the same day.[52]

Examples of the month being held as complete, even though it is only begun, are not as numerous as those of the year, but they are nevertheless found in the works of the canonists. Thus Engel (†1674) refers to the 7th month as beginning with the 181st day when the legitimacy of one's birth is involved.[53]

This lack of precision concerning the length of the last year or month of any given amount of years or months still existed prior to the Code. One general application only seems to have been mentioned by the authors toward the end of the last century, and that was in favorable matters when the law was not clear.[54] The best example in recent years was perhaps the case of the age required for the reception of the major orders. The Council of Trent declared:

[52] "Cum annus designatus ad susceptionem cujuscunque ordinis, inceptus habetur pro completo, non licet ordinem suscipere in mane diei, in cujus vespere quis complet aetatem requisitam, quia non dicitur annum attigisse sequentem, qui antecedentem etiam unius horae spatio nondum compleverit . . . "—Ferraris, *Bibliotheca,* s.v. *Aetas,* n. 51.—C. D'Annibale, *Summula,* I, n. 39, note 34; St. Alphonsus, *Theologia Moralis,* lib. VI, n. 800, where mention is made of Diana and the Salmanticenses as holding a different opinion on the grounds that a little matter is reputed for nothing, or because a day begun is considered as completed in favorable matters, or again because morning and evening are normally considered the same day.

[53] *Collegium,* lib. IV, tit. XVII, n. 6.

[54] Cf. D'Annibale, *Summula,* I, n. 39: Maroto, *Institutiones,* I, n. 257, 5, note 2.

> No one shall for the future be promoted to the order of subdeaconship before the twenty-second year of his age; to that of deaconship before his twenty-third year; to that of the priesthood before his twenty-fifth year.[55]

§3—The *Dies Terminus ad Quem*

The time has now come to consider the cases when the year and the month, or any other period of time such as one week, 15 days, etc., must be actually completed. Thus, if the law prescribed the completion of the 20th year, one had to know whether or not that occurred with the beginning or with the end of the last day of that year.

Various theories were proposed according to which the last day was to be counted in certain instances and to be omitted in others. The most plausible of these explanations, and the only one to have endured well nigh to the Code, held that the last day when once begun was to be considered as a complete day in favorable matters, but not in odious affairs.[56]

Antonellus mentioned one theory according to which the natural reckoning was used when a time-interval was specified for the loss of a benefice. On the other hand when a time-interval led to the acquisition of a benefice it was then sufficient to reach the last day. But at the time of Antonellus this theory had been discarded in the sense that the authors no longer mentioned specifically the term benefice. The common teaching simply held that a day begun was to be looked upon as a complete day in favorable but not in odious affairs.

Consequently a marriage could be validly contracted when the groom had reached the last day of his 14th year, and the bride the last day of her 12th year. Similarly it was sufficient to reach the last day of the prescribed time-interval in the *usucapio* and

[55] Sess. XXIII, *de ref.* c. 12.—Cicognani, *Canon Law,* p. 683. The Code uses the numbers 21, 22, and 24 instead of 22, 23, and 25, but in such a way as to mean exactly the same thing as the Council of Trent. Canon 975 states: "Subdiaconatus ne conferatur ante annum vicesimum primum completum; diaconatus ante vicesimum secundum completum; presbyteratus ante vicesimum quartum completum."

[56] Cf. D'Annibale, *Summula,* I, n. 39.

the *praescriptio longi temporis,* because these were among the *favorabilia.* However, this did not apply to the prescription of 30 years since it was considered as odious.[57]

D'Annibale referred to the case of the novice who would make his profession on the last day of his 16th year, and to that of the minor who would assume an obligation on the last day of his 21st year. Both these acts would be invalid because they impose an obligation and are therefore considered odious things in law. Consequently, the last day of the time-intervals under present consideration had to be fully completed before a valid act could be posited.[58]

Strange consequences often followed from the above mentioned interpretation. Let us take for example a certain contract wherein a sum of money was to be paid within a year. As soon as the last day of the year was begun the creditor could lawfully demand the sum of money. The debtor, however, was not obliged to surrender said sum before the end of the last day.[59]

Toward the close of the last century the notion of the term at which a time-period stood as completed changed to a certain extent. Many authors of note insisted, at least for a few institutes, matrimony for instance, that when complete years or months were mentioned, the last day also should be physically completed.[60] Gasparri, for instance, insisted that the last day

[57] Cf. *De Tempore Legali,* lib. IV, c. I, nn. 1-6. The distinction between favorable and odious matters was upheld by Barbosa, *Tractatus Varii* (Lugduni, 1660), I, *De Axiomatibus Juris Usufrequentioribus,* Axioma 71, n. 6; Benedictus XIV, *Quaestiones Canonicae,* q. 197—Wernz, *Ius Decretalium,* IV, 319; Pirhing, lib. I, tit. XI, n. 1, not. 12; Reiffenstuel, lib. IV, tit. II, n. 13; Sanchez, *De Matrimonio,* lib. VII, disp. CIV, nn. 2-3.

[58] *Summula,* I, n. 39, note 35.

[59] Cf. Ojetti, *Synopsis,* s.v. *Tempus.*

[60] Cf. Maroto, *Institutiones,* I, n. 257, 5, note 2. It is here to be noted that even when the old distinction between favorable and odious matters was retained, authors did not agree as to the nature of what was favorable. In other words, the question of the last day was often known to vary according to the nature of the canonical institute involved. An application of this difference was found in the works concerning matri-

had to be completed physically, at least for marriages to be contracted. If marriages had been previously entered into before the completion of the last day of the prescribed age, these unions were to be tolerated unless the Holy See decided otherwise.[61] Wernz adhered to a similar opinion more or less hesitatingly.[62]

This author, however, made no such distinction (*contractis-contrahendis*) when he spoke of the complete years required for the reception of the tonsure and the episcopate. He simply stated

mony and betrothal. According to Pirhing (lib. IV, tit. II, n. 2, not. III) the beginning of the last day was sufficient for either institute.—Schmalzgrueber, on the other hand, distinguished between the two. The beginning of the last day of the prescribed age sufficed for marriage, but not for a valid betrothal: "Non tamen debet aetas ista, et annus esse completus omnino physice, sed sufficit, si moraliter; quo differt matrimonium a sponsalibus; in matrimonio enim propter specialem illius favorem ultima dies inchoata pro completa habetur."—Lib. IV, tit. II, n. 60.

[61] *Tractatus Canonicus de Matrimonio* (2 vols., Paris, 1891), n. 495. This work will hereafter be cited as *De Matrimonio.*

[62] "Quod septennium ad sponsalia requisitum, *saltem si non constat malitiam supplere aetatem,* non tantum *moraliter,* sed *vere completum* sit necesse est. Nam in cap. 4. 5. X. h. t. adhibentur verba *"complesset," "compleverat."*

"Quare si auctores quidam putant *legi* satisfactum esse, etiamsi *integer dies,* duo vel tres dies, integra hebdomada, imo sex menses desint, haec ipsa varietas opinionum satis probat computationem illam *moralem* niti arbitraria fictione. Porro si lex ex *iuris* praesumptione praescribit *certum tempus,* praesumptio *privata* non habet locum, et verba legis accipienda sunt sensu *proprio* et in *iure* recepto, sed annus sensu proprio et iuridico non est completus, si desit integer dies, hebdomada, etc. Quodsi complures canonistae annos infantiae completos esse existimant, si ex computatione quadam iuris Romani . . . *ultima* dies septennii sit *inchoata,* haec sententia est solide probabilis. Nam Ecclesia generatim in huiusmodi rebus v.g. de domicilio, de aetate *adoptavit ius Romanum,* nisi suo iure speciali de contrario caverit v.g. de anno novitiatus; lex autem *specialis* Ecclesiae in contrarium hac in re non videtur existere, imo Ecclesia nos pari. ratione remittit ad *leges* i.e. Romanas et ad *canones* v.g. Nicol. I, cap. 2, X. h.t. At si agatur de sponsalibus non *contractis,* sed *contrahendis,* certe ex disciplina vigente altera sententia est *tutior,* quae requirit aetatem etiam quoad *ultimum* diem *physice* completum."—*Ius Decretalium,* IV, n. 314. Cf. also *ibid.,* n. 319.

that it was probable that the age required was reached with the beginning of the last day of the last year prescribed.[63]

These fluctuations in the application of the civil reckoning led many authors to favor the natural computation for the reckoning of the year prescribed for the novitiate. Thus Reiffentuel, while admitting the diversity of opinions in the matter, maintained that the natural reckoning was more commonly advocated by the canonists, was in stricter conformity with law and reason, and was consequently to be used in practice. It was furthermore held that a profession made half an hour or even fifteen minutes before the expiration of the prescribed time would not be valid.[64] It must be noted, however, that the civil reckoning was officially prescribed a few years before the promulgation of the Code.[65]

§4—The *Dies Terminus a Quo*

According to Van Hove authors have maintained divergent opinions concerning the *dies a quo.* The general trend, however, has always been not to consider that day as a full day whenever obligations were involved. In favorable matters writers kept the adage: *"Dies (termini a quo) incoeptus habetur pro completo."*[66]

It appears that in practice the wording of the law, and not the nature, favorable or otherwise, of the institute involved was the determining factor. That is clearly shown from the writings of Antonellus, who gave the matter ample treatment.[67]

Article 4—Physical or Moral Reckoning

Before trying to say whether or not physical or moral reckoning was used one must first of all define one's terms and then inquire whether or not these have had the same meaning throughout the history of the Church.

[63] *Op. cit.*, II, n. 113.

[64] Lib. III, tit. XXXI, n. 94. Cf. Sanchez, *De Matrimonio,* lib. II, disp. XXIV, n. 22; Conc. Trident., sess. XXV, *de regularibus,* c. 15.

[65] S. C. de Religiosis, decr. 3 maii 1914—*Fontes,* n. 4419.

[66] *De Tempore Supputatione,* n. 274. Cf. Barbosa, *De Axiomatibus Iuris,* Axioma 71, n. 5; Engel, *Collegium,* lib. I, tit. IV, n. 11; De Angelis, *Praelectiones,* lib. IV, tit. VIII, n. 9.

[67] Cf. *De Tempore Legali,* lib. IV, c. II, nn. 1-19, where the legal import of various expressions and particles is examined.

Today the word *moral,* as applied to time-reckoning, simply signifies a more or less accurate computation which at least approximates the reality of the truly physical reckoning of time. Thus a moral interpretation or reckoning of midnight would allow one, when fasting from midnight on, to take food even 10 or 15 minutes after the moment which makes midnight according to any of the physical standards indicated in canon 33, § 1. A concrete example of a moral reckoning of time is afforded by Cappello, who at one time taught that the month mentioned in canon 858, § 2, could be so interpreted as meaning but 26 or 27 days.[68]

The reckoning of time on the other hand is said to be *physical* when no such latitude is given to the individual. Physical reckoning simply does away with the rule "a very small matter is held as nothing." [69]

But, *physical* is also opposed to *mathematical* in time-reckoning. This latter form of computation would consider only the exact motion of the celestial bodies and rely on cold arithmetical computations as one's guide. Time is said to be reckoned physically, as opposed to mathematically, when one makes a good clock his guide and then abstracts from any possible deviation to which this instrument might be subject. The readings of the clock are then final and allow of no latitude, such as the moral reckoning would.

This mathematical reckoning of time was never looked upon as a canonical entity. For, as La Croix (†1714) explained, it was impossible to reckon time according to the exact motion of the sun. Time was to be reckoned according to the manner approved by the Church, that is, by following a good clock as one's guide.[70]

If one now casts a glance in retrospect through the history of Canon Law one finds that the authors who favored a moral

[68] *Tractatus Canonico Moralis de Sacramentis,* 3 vols. in 4 tomes, Vol. I (Romae: Marietti, 1921), n. 506. Cf. *infra,* p. 244, where this question is more thoroughly examined.

[69] Cf. Cicognani, *Canon Law.* p. 682.

[70] *Theologia Moralis,* lib. I, n. 584. The words *physically* and *mathematically* are used in direct contraposition by Cicognani (*Canon Law,* p. 684).

reckoning of time at least for certain matters have always been a small minority, and that this opinion had been altogether discarded for many a year before the Code. D'Annibale, for instance, spoke without the least doubt on this point. He stated that in the computation of time there was no question of using the rule: *minimum pro nihilo reputatur.* And the reason given was that in those things which are defined by nature or law no moral estimation could be used. Therefore, the smallest moment of time was to be computed.[71]

The word *moral* has at times been used in a slightly different sense. One often reads here and there that the interval of time to be observed between the reception of the various orders is to be reckoned, not metaphysically, but morally. Thus, the year that must elapse between the last of the minor orders and the subdiaconate is said to be reckoned morally, that is, not according to the year of 365 days, but according to the liturgical calendar, for instance, from Trinity Sunday to Trinity Sunday, even though there be as few as 340 days in a given year.[72]

Article 5—The Day and the Hours

The general reckoning of the year and the month, vague and somewhat indeterminate as it is, looks like a model of precision when compared to that of the day. In the first place the *dies* could be used in law to express any length of time.[73]

The more usual acceptation, however, of that word referred to a period of 24 hours to be reckoned from midnight to midnight or

[71] *Summula,* I, n. 39. The same opinion was held by the following authors: Engel, *Collegium,* lib. I, tit. XIV, n. 5; De Justis, *De Dispensationibus Matrimonialibus* (Lucae, 1727), lib. III. c. VIII, n. 17; La Croix, *Theologia Moralis,* lib. I, n. 582; St. Alphonsus, *Theologia Moralis,* lib. II, n. 282; Schmalzgrueber, lib. IV, tit. II, n. 60; Reiffenstuel, lib. III, tit. XXXI, n. 94; Sanchez, *De Matrimonio,* lib. VII, disp. CIV, n. 23, 3.

St. Alphonsus (*Op. cit.,* lib. VI, n. 800) mentions Diana and the Salmanticenses as holding a divergent view.

[72] Cf. Bonacina, *Opera Omnia* (3 vols., Venetiis, 1687), *De Censuris,* disp. III, q. I, punctum IV, #11, n. 2; Engel, *Collegium,* lib. I, tit. XI, n. 20; Antonellus, *De Tempore Legali,* lib. I, c. III, n. 10.

[73] Cf. Ojetti, *Synopsis,* s.v. *Tempus.*

from any given moment of one calendar day to the corresponding moment of the next calendar day.[74]

At one time an outstanding exception to this rule was the reckoning of the day in odious affairs. The natural day (from dawn to dusk) was, according to some authors, the accepted interpretation.[75]

This is rather exceptional and is not mentioned by writers of a later period. The great problem to solve concerning the day is to find the starting point. An effort was made in Article 3, §1, to show in which cases the 24 hours were reckoned from moment to moment. In all other instances two fixed or legal starting points have been used in the Church: midnight, according to the Roman style, and sunset according to the Hebrew reckoning.

One may affirm without the slightest doubt that the Jewish reckoning was used from the very beginning of the Church in liturgical and ceremonial matters, whereas the Roman concept of the day came along with the other institutes when the Church in a large measure borrowed from the Roman legal system. Definite knowledge concerning the precise use of the two reckonings during the first thirteen centuries of the Church's history does not seem to have been acquired. Little is known beyond the fact that the general tendency in this matter has always been to drop the Jewish system in favor of the Roman. This evolution has wiped out the sunset to sunset day, even to the extent that it is not

[74] Cf. Antonellus, *De Tempore Legali,* lib. I, c. III, nn. 27-38; Baldus, *In Decretalium Volumen Commentaria* (Venetiis, 1580), *super* II *Decretalium, De Feriis,* cap. 2; Gonzalez-Tellez, *Commentaria Perpetua,* lib. I, tit. XXIX, c. XXIV, n. 5.

[75] "Notandum tamen est, quod ubi in dispositione legis, vel hominis sit mentio de die, in favorabililbus intelligitur de die naturali: in odiosis vero, et poenalibus intelligitur de artificiali, non de naturali. Unde si statutum puniat gravius delinquentem in die S. Laurentii, huiusmodi poena non haberet locum contra delinquentem in nocte praecedente, vel subsequente dictam festivitatem."—Antonellus, *De Tempore Legali,* lib. 1, c. III, n. 34. It must here be recalled that Antonellus defined the natural day as the period of time between two successive midnights, and the artificial day as the period of light between dawn and dusk. Cf. *supra,* p. 73.

even used today in many liturgical institutes in the strict acceptation of the word.[76]

For a number of years the reckoning of the 24 hours for the sanctification of the Lord's day, as also for the fulfilling of the precept of fast, were reckoned from sunset to sunset. The same computation also applied to the gaining of indulgences and was still in use before the promulgation of the Code.[77]

Thus one reads in the *Decretum Gratiani* that the faithful are to abstain from work every Sunday throughout the year from sunset to sunset.[78] However, when the *Corpus* was commented upon the above mentioned reckoning was actually reserved to purely liturgical affairs.[79] The same doctrine is also found in the Decretals.[80]

The only conclusion to be drawn from the above texts is that the law itself stipulated that the reckoning of the 24 hours in view of the abstention from servile works was to have as starting point sunset or dusk. Custom intervened to introduce the midnight to midnight reckoning. No precise time is given by the authors for

[76] Thus the private recitation of the breviary may be begun at 2 P.M. of the previous calendar day. The obligation itself is reckoned from midnight to midnight. Cf. canon 923 for the reckoning of time in the matter of gaining indulgences.

[77] Cf. Benedictus XIV, litt. encycl. *"Peregrinantes,"* 5 maii 1749—*Fontes*, n. 399. In #6 the ecclesiastical or "sunset to sunset" day is clearly determined: "Qui . . . Basilicas semel saltem in die, per triginta continuos, aut intervallos dies, sive naturales, sive etiam Ecclesiasticos, nimirum a primis vesperis unius diei usque ad integrum ipsius subsequentis diei vespertinum crepusculum computandos . . . devote visitaverint . . . "

[78] "Pronuntiandum est laicis, ut sciant tempora feriandi per annum, idest omnem Dominicam a vespera usque ad vesperam."—C. 1, D. III, *de cons.* Cf. also c. 5, D. LXXV.

[79] "Sed tamen no [ta] quod inceptio ista non habet locum quoad omnia. Nam secundum hoc saepe comederemus carnes in die veneris. Unde dic quod quoad solutiones faciendas, quoad carnes comedendas, dies incipit a media nocte, et terminatur in aliam mediam noctem, et maxime quia hoc dicit lex [i.e. Romana], et hoc habet consuetudo, licet quoad solutiones distinguant quidam inter clericos et laicos: quod non approbo."—Glossa to *usque ad vesperam* (C. 1, D. III, *de cons.*)

[80] Cf. C. 1, X, *de feriis*, II, 9. Cf. also Cc. 2, 5, *ibidem.*

this change.[81] The common teaching at least in the 17th century, held that custom was everywhere to be followed, but that in Europe the midnight to midnight reckoning was the accepted one as designated by customary usage.[82]

A slightly different starting point was also mentioned in the sources for the observance of the law of abstinence. Leo IV (†855) approved the custom according to which the reckoning of the day in view of the observance of the law of abstinence was made from the end of the evening meal to the following sunset.[83]

Tagle states that the sunset to sunset reckoning was even used in purely juridical matters.[84] He argues from the famous document that Innocent III sent to the Archbishop of Toledo.[85] It is true that nightfall was here mentioned, but there was no question of reckoning the day from sunset to sunset. The document simply stated that certain juridical acts were not to be posited after nightfall.[86] It is furthermore highly improbable that

[81] Cf. Panormitanus, *De feriis,* c. II, n. 3; Antonellus, *De Tempore Legali,* lib. I, c. III, n. 38; Benedictus XIV, *Quaestiones Canonicae,* q. 181; Engel, *Collegium,* lib. II, tit. IX, n. 14; Laymann, *Theologia Moralis,* lib. IV, tract. VII, c. II, n. 1.

[82] Cf. Laymann, *Theologia Moralis,* lib. IV, tract. VII, c. II, n. 1, where reference is made to many authors.

[83] "De esu carnium apud vos vetustissima, et non improbanda traditio semper est tenenda, ut a coenae termino, quae fit in principio noctis quartae feria, quae lucescit in quinta feria, usque in diluculum quintae feriae . . . "—C. 11, D. III, *de cons.*

[84] *De Temporis Supputatione,* p. 28.

[85] "Ne vero iudicium ecclesiasticum exerceatur in tenebris, quia iuxta testimonium Veritatis qui male agit odit lucem, licet etiam secundum consuetudinem et constitutiones legitimas more Romano dies a medietate noctis incipiat, et in medio noctis desinat subsequentis, unde quod in duabus dimidiatis noctibus agitur, perinde sanxit habendum legalis auctoritas, ac si qualibet parte lucis agatur, nos tamen eam procedendi horam congruam intelligimus, ex quo possis ante noctis tenebras perficere, quod incumbit."—C. 24, X, *de officio et potestate iudicis delegati,* I, 29.

[86] In Roman Law a similar prohibition existed for the judicial sentence which could not be given after dark. Cf. Gonzalez-Tellez, *Commentaria Perpetua,* lib. I, tit. XXIX, c. XXIV, n. 5.

the Church should have used the Jewish reckoning of the day along with the other computations taken from the Roman Law. At any rate no mention of this is to be found either in the sources or in the authors.

To summarize one may say that for the past seven or eight centuries in the known parts of the world the reckoning of the day in disciplinary affairs has been exclusively from midnight to midnight. Strange to say, this particular point is the only time-institute on which the Church has legislated in a general way. In spite of this doubts and contradictions have not vanished, not even with the Code. Authors seem to have taken a morbid delight in "interpreting" clear legal texts. At times contradictory norms for similar cases emanated from the Holy See.[87]

Two periods can be distinguished here. The first goes up to the middle of the nineteenth century and is characterized by the use of a none too secure mode of reckoning midnight or any other starting point for the computation of the 24 hours, sunset or dusk for instance, and above all, by the fact that only one official time was used, namely local time, which was in full accordance with the usual reckoning. The other phase of the problem began with the introduction of legal or regional time.

During the first period the only difficulty was the proper determination of the one and only starting point with the more or less primitive instruments at one's disposal. This inconvenience, decreasing as it did with the years, did not totally disappear, but was, in the second period, overshadowed by the problem whether or not regional time supplanted local time, whether or not they both could be used disjunctively. In other words, the problem was exclusively a physical one in the first period. In the second, it became mostly canonical.[88]

The first known decision of the Holy See relative to the deter-

[87] Cf. *infra*, pp. 94-95.

[88] The problem of the physical determination of any given starting point, midnight or dusk, is not directly a canonical question and will consequently not be given separate treatment. It will be casually mentioned in the next article.

mination of a starting point for the reckoning of the 24 hours of the day clearly indicates the confused notions prevalent at the time. It was asked whether clerics living within the Arctic circle could use the meridian of Rome for the computation of time in the recitation of Lauds. The decision was based on Fr. Secchi's *votum.* This illustrious astronomer stated that the distribution of the hours of the day should be according to the Roman method, but that the reckoning of noon should be determined by the transit of the sun over the local meridian. To obtain mean time one had to take into consideration the equation of time.[89]

The first decision emanating from the Holy See and approving the use of a non-local time was given by the Sacred Penitentiary to the Archbishop of Naples, June 18, 1873. The occasion for this decision was the adoption by Naples of the mean time of Rome. The answer stipulated that the faithful of Naples could but were not obliged, to follow Rome mean time for the observance of the natural fast and other ecclesiastical obligations (per l'adempimento del digiuno naturale ed altri obblighi ecclesiastici).[90]

The text obviously meant that one was then free to use either Rome mean time or Naples mean time. In spite of that the decision was given a different interpretation by a few authors.[91] For instance, the commentator of the above mentioned decision in the *Acta Sanctae Sedis* maintained that the faithful in Naples could not use their own local time if the public clocks indicated a different time.[92] In other words one had to follow the public clocks of one's community.

[89] S. C. Rituum, *Missionis Poli Artici,* 6 Februarii 1858—*ASS,* III (1867), 602-603. Cf. *ASS,* VII (1872), 400, note 1.

[90] S. Poenit., 18 iun. 1873—*Fontes,* n. 6430.

[91] Cf. Van Hove, *De Temporis Supputatione,* n. 276, note 4.

[92] "Ex quo responso colliges:

1. Publicis horologiis signantibus tempus medium (quod in parte anni praecedit tempus verum, alia anni parte subsequitur) quamquam non desint privata horologia quae tempus verum signent [?], fideles sequi debere, causa ieiunii naturalis et ecclesiastici aliorumque Ecclesiae officiorum, publica horologia licet tempus verum non signent.

Other writers, however, understood the document in its true meaning. According to them the faithful of Naples were free to use either Rome mean time or Naples mean time.[93]

To complicate matters the Sacred Congregation of Rites gave a different solution to a case which was similar in character to the preceding one. It seemed to confirm the opinion set forth in the *ASS*.[94]

It was asked whether, for the reckoning of time in the obser-

2. Haec locum quoque habere si tempus medium signent non proprii meridiani sed alieni, quamquam hoc tempus magis discrepet a tempore vero.

3. In temporis enim designatione eam regulam Ecclesia sequitur, quae omnia hominum negotia in singulis locis publice dirigit."—*ASS*, VII (1872), 399-400. The same opinion is held by D'Annibale (*Summula*, I, n 39, note 28)—It must be borne in mind that the above mentioned interpretation was not the *votum* of a consultor when the question was still in process of being solved, but was really the expression and teaching of a canonist when the response had already been delivered and sent to Naples.

It is here to be noted that the authors did not all give the same meaning to the word true time—*tempus verum*. The true time referred to in the above quotation—*quamquam non desint privata horologia quae tempus verum signent*—cannot easily be looked upon as the same true time mentioned on p. 43 of this work, for the simple reason that no clock has ever been known to keep up with the sun's vagaries. However, when the commentator spoke of private clocks *(privata horologia)*, he possibly referred to the sun-dial. Otherwise he must have had in mind the old Italian method of counting the hours of the day from sunset to sunset and not from midnight to midnight. In the old Italian system the clocks were set according to the actual setting of the sun. This starting point that varied with the actual motion of the celestial bodies could be called true as opposed to a mean or fictitious sunset that occurred every day at 6 P. M.

[93] Cf. Gennari, *Consultations de Morale, de Droit Canonique et de Liturgie* (Translated from the Italian by Boudinhon, 5 vols., Paris, 1907-1910), Vol. V, *Liturgie*, Consultation 47, n. 8. This work will hereafter be referred to as *Consultations*.

[94] Cf. Creusen, "Minuit Canonique" ou "Loi Pure et Simple,"—*Nouvelle Revue Théologique*, L (1923), 464-474. This article will hereafter be referred to as *Minuit Canonique*, and the letters *NRT* will be used to cite the above mentioned periodical.

vance of various ecclesiastical obligations, one could follow mean time. The reply was: *"Standum publicis horologiis"*.[95]

A few years later the Sacred Penitentiary gave a new decision confirming the one of June 18, 1873. The question was thus stated: "Utrum, ubi horologia adhibentur, tempori medio accomodata, ipsis sit standum, tum pro onere divini officii solvendo, tum pro ieiunio naturali servando; vel debeat quis, aut saltem possit uti tempore vero?"—The reply was: "Fideles in ieiunio naturali servando, et in officio recitando, sequi tempus medium posse, sed non teneri." [96]

According to Many the Sacred Penitentiary considered the decision of the Sacred Congregation of Rites as binding only in the diocese for which it was given.[97]

At any rate the Sacred Congregation of Rites soon modified its opinion. In 1898 work was begun on an authentic edition of the decrees of this Congregation, and, as Van Hove remarks, a few changes were made here and there.[98] The decree in question was given under its old number in the previous collections but the answer was now: "Posse stare publicis horologiis." [99] In 1905 the same Congregation immediately answered *Ad libitum* to the question whether, for the recitation of Matins, one could use indiscriminately either Greenwich or local time.[100]

In the meantime the Sacred Congregation of the Council had rendered a decision similar to those given by the Sacred

[95] S. R. C., *Clodien.*, 7 aug. 1875—Gardellini, *Decreta Authentica Congregationis Sacrorum Rituum* (3. ed., 5 vols., Romae, 1856-1879), V, Appendix IV, n. 5622. This work will hereafter be cited as Gardellini.

[96] S. Poenit., 29 nov. 1882—*Fontes*, n. 6434. It must not be forgotten that the words mean and true *(medium, verum)* did not, in 1882, necessarily have the same meaning as they now have. *Medium* often meant what is now called regional, and *verum* simply corresponded to local. Cf. *infra*, pp. 97-99.

[97] *Praelectiones de Missa* (Paris, 1903), n. 21, 1. This work will hereafter be referred to as *De Missa*.

[98] *De Temporis Supputatione*, n. 276, note 5.

[99] *Decreta Authentica C. S. Rituum* (7 vols., Romae, 1898-1927), III, 58—n. 3365 Clodien—(n. 5622).

[100] S. R. C., *Placentia in Hispania*, 12 maii 1905—*Fontes*, n. 6338.

Penitentiary. It stated that legal time could be followed.[101]

In spite of all these decisions there still remained many problems to be solved by the canonists. According to the first three decisions given by the Holy See (S. Poenit., 18 iun. 1873; S.R.C., *Clodien.*, 7 aug. 1875; S. Poenit., 29 nov. 1882) one thing was certain, although it continued to be denied by some—and that was that regional time could certainly be used where the public clocks indicated it. On the other hand, authors of note doubted whether the opposite was to be accepted. In other words, could one use regional time where the public clocks showed local time? Gennari held the opinion that regional time could not then be used. He argued that the Church prescribed as a rule the use of the time adopted for public affairs in the various communities, that is, local time. If the Church permitted the use of mean (i.e. regional) time, this was done simply for the purpose of obviating doubt and anxiety among the faithful.[102]

Gennari's terminology clearly shows the complicated status of the question. In the first place he used the words *medium* and *verum* in such a way as to make them correspond to regional and local.[103]

[101] S. C. C., *Trevir,* 22 iul. 1893—*Fontes,* n. 4287.

[102] "Car la raison qui porte le Saint Siège à permettre de suivre les horloges publiques, quoique réglées sur le temps moyen, est, ce semble, le souci de ne pas faire surgir des anxiétés et des troubles parmi le peuple, or cet inconvénient n'est pas à craindre quand les horloges ne sont pas communément réglées sur le temps moyen. C'est pourquoi Del Vecchio (ap. Scavini, t. II, n. 194) dit: Ecclesia in temporis designatione eam regulam sequi solet, quae omnia hominum negotia in singulis locis publice dirigit. D'autre part, s'il était libre à chacun de suivre toujours partout le temps moyen, comme la différence avec le temps vrai est considérable dans les pays éloignés, il s'en suivrait que l'on arriverait à faire de jour ce qui doit être fait de nuit, et réciproquement; et il n'est personne qui ne voie combien la chose serait étrange."—*Consultations,* V, cons. 47, n. 9.

[103] That is clearly indicated in the preceding note. Cf. *op. cit.*, V. cons. 47: "Ainsi l'illustre P. Secchi estimait que les missionnaires des régions arctiques pouvaient diviser les heures suivant la méthode romaine; mais que pour les compter, ils devaient suivre le méridien local, c'est-à-dire le temps vrai." Cf. also the decision of the Holy Office of May 9, 1892.

In the second place the meaning of the word *medium* does not correspond adequately to our modern concept of regional. Gennari did not seem to realize the possibility of zone time. To him mean time was Rome mean time and nothing else. If he had visualized any other mean or regional time he would never have come to the conclusion that the use of mean time in the remote regions of the world would change day into night and vice versa.

One might now wonder whether it was lawful, then, to use true time according to the present day acceptation of the word. That is doubtful. In the first place it was never indicated or mentioned in the decisions of the Holy See. Many, who, incidentally, clearly defined true time according to our way of looking at it, held that it could not be used.[104] Lehmkuhl, however, stated that it might be used. He regarded it as lawful at one's own discretion and preference to follow either true sun time, or mean time, or also the time employed in public by the community, even if the latter reckoning of time did not agree with either of the two previous ones.[105]

The expression *tempus zonarium* is found for the first time in official documents in a decision of the Holy See given August 9, 1892. The Archbishop of Utrecht, April 27, 1892, proposed the following doubt to the Holy Office:

> Quandoquidem a Ia die mensis Maii 1892 omnia horologia viarum ferrearum per totam Neerlandiam in indicandis horis regulam sument tempus medium loci Greenwich in Anglia, quod tempus tertiam fere horae partem retro distat a medio tempore in Neerlandia, diciturque *tempus zonarium,* Gubernium civile praescripsit ut et ipsa horologia officiorum publicorum expediendis lit-

Greenwich mean time is opposed to *verum tempus iuxta meridianum proprii loci.*

[104] *De Missa,* n. 21, 1.

[105] *Theologia Moralis* (12. ed., 2 vols., Friburgi Brisgoviae, 1914), II, n. 216, 2. This is another instance of the discrepancy between the decisions of the Holy See and the interpretations of the authors. True or apparent sun time was not mentioned in the sources as such before the Code.

> teris nuntiisque telegraphicis idem tempus medium loci Greenwich indicarent. Quia *insuper,* sive a magistratu civili, sive per usum, alia quoque horologia publica tempori praefato *multis saltem in locis* conformabantur, quaestio exsurget: Utrum possint clerici et fideles per totam Neerlandiam in ieiunio naturali, ceterisque ecclesiasticis obligationibus observandis, observare tempus medium loci Greenwich, an vero sequi debeant verum tempus iuxta meridianum proprii loci.

On May 9, 1892, the answer was given: *"Affirmative ad primam, negative ad secundam partem."* In other words, Greenwich mean time could be used in Holland. In spite of this the Holy Office had to intervene a second time to ease a few doubtful consciences.[106]

It was now beyond doubt that the faithful in Holland could use Greenwich mean time even in those communities where the public clocks indicated local time or any other mean time different from that of Greenwich. The only explanation of this is that Greenwich mean time had been prescribed by the Government for official use by the railroads and the telegraph offices. Indeed, Greenwich mean time, it was stated, could be used throughout Holland in spite of the fact that in only a few places the public clocks had adopted said time.[107]

[106] "Nunc vero aliqui iterum dubitant, contenduntque, clericos et fideles in Neerlandia pro ieiunio naturali aliisque ecclesiasticis praeceptis observandis uti non posse praedicto medio tempore Greenwich, eo quod de facto alia horologia publica non multis in locis praefato tempori se conformarunt. Cum autem pro ieiunio in primis naturali servando res non sit levis momenti, dignetur Sanctitas Sua benigne sequens solvere dubium:

Utrum, non obstante quod de facto non multis in locis alia horologia publica se conformarunt tempori medio Greenwich, clericis et fidelibus in Neerlandia in ieiunio naturali ceterisque ecclesiasticis obligationibus servandis, licitum sit sequi tempus illud medium Greenwich, quod inde a Ia die mensis Maii 1892 per totam Neerlandiam in omnibus officiis publicis tam viarum ferrearum, quam litterarum et telegrammatum expediendorum introductum fuit?"

An answer was given in the affirmative on August 9, 1899. This decision was approved two days later by Leo XIII.—*ASS,* XXXII (1899-1900), 250-252.

[107] Cf. Michiels, *Normae Generales Juris Canonici* (2 vols., Lublin: Universitas Catholica, 1929), II, 139.

The logical conclusion to draw from all these decisions of the Holy See was that a person was allowed to choose, *disjunctive* at least, any one of the various reckonings now explicitly granted by canon 33, § 1.[108] The possibility of using two or more systems of time-reckoning for the fixation of the same day for various obligations was not considered by the authors before the Code with the exception perhaps of Vermeersch. The classical question concerning one's line of conduct referred to the case when two or more clocks—all of which were supposed to give one and the same time—differed.[109]

There is, however, one fundamental difference between canon 33, § 1, and the legislation that preceded the Code. Today, as shall be shown in the commentary, authors are not agreed upon the enumeration given in the above mentioned canon. Many hold that the privilege of using any one of a number of recognizable points of time as indicative of the moment of midnight is restricted to the cases explicitly mentioned. Others, on the contrary, are of the opinion that the enumeration of the cases mentioned by canon 33, § 1, is not meant to be exclusive with regard to other similar cases, but rather is intended to be illustrative of them. The privilege would then apply to other precepts of a private nature such as the abstaining from servile works on Sundays and holy days of obligation. Whatever may be said of the present day law, there was absolutely no doubt as to the number of precepts that fell under the decisions of the Holy See prior to the Code. All ecclesiastical precepts without exception were included. The words *ceterisque ecclesiasticis obligationibus* or their equivalent are found more than once.[110]

[108] The use of zone time (extraordinary), regional time(e.g., Roman mean time for Naples), and local time (mean) had been explicitly conceded by the Holy See. True time was included by the authors.

[109] "Si étrange qui cela paraisse, nous ne trouvons la question *ainsi posée* (avant le Code) que dans un cours lithographié du R. P. Vermeersch. Les autres auteurs confondent cette question avec une autre, toute différente: 'Puis-je, pour diverses obligations, suivre en même temps deux horloges qui marquent des heures différentes?' "—Creusen, "Minuit Canonique,"—*NRT*, L (1923), 468.

[110] Cf. S. Poenit., 18 iun. 1873—*Fontes*, n. 6430; S. C. S. Off., 9 aug 1899—*ASS*, XXXII (1899-1900), 251-252.

There is another problem which is closely associated with the one under present consideration. When the clocks began to strike the hours one could wonder whether the first or the last stroke determined the end of the preceding hour. The teaching of the authors has been fairly uniform in this respect. The common opinion has held from the very beginning that the first stroke marked the end of the preceding hour.[111]

Article 6—Dawn and Dusk

Before the Code the period of time wherein Mass could be celebrated ran from dawn to noon.[112] By dawn (*aurora*) was meant the entire period of time which was intermediate between complete darkness and sunrise.[113] A more precise definition was given by Many. Dawn was said to be that diffusive light that precedes the rising of the sun. The cause of this light was attributed to the sun's rays which illumined the upper regions of the sky just before the sun reached the horizon. The period of time that lapsed between the first moments of this illumination and the actual rising of the sun was called dawn, and corresponded to dusk which was the period of time that intervened between the setting of the sun and complete darkness. The duration of dawn depended on the latitude of one's locality, and varied, in any given place, throughout the year.[114]

[111] Cf. D'Annibale, *Summula,* I, n. 39; Schmalzgrueber, lib. III, tit. V, n. 306. Galileo, in 1610, discovered the invariability of pendulum oscillations. The application of that principle to the control of clocks was made by Huyghens in 1656. Cf. Mitchell, "What Time is it? Midnight and Fasting."—*ER,* LXXXIV (1931), 491-500, especially p. 493.

[112] "Missa privata, saltem post matutinum et laudes, *quacumque hora ab aurora usque ad meridiem dici potest.*"—Rubricae Missalis, P. I, tit. xv, n. 1.

[113] Cf. Ojetti, *Normae Generales,* p. 198, notes 3 and 4.

[114] *De Missa,* n. 20, 2, a—The scientific explanation of the actual beginning of dawn was thus given by the same author: "Incipit aurora seu diluculum, quando sol jam *decimo octavo gradu* proximus est horizonti; sicut desinit crepusculum serotinum, quando sol jam recessit ab horizonte ultra *decimum octavum gradum.*" *Op. cit.,* n. 20, note 1.

In practice, however, the beginning and the end of the natural day

The determination of the time when one could begin Mass was relatively simple for those countries where dawn and dusk actually occur every day of the year. A practical problem, however, had to be solved when missioners began to preach the Gospel within the Arctic circle. There the word *aurora* had no meaning. It is not surprising to find, then, that efforts were made to change the wording of title xv of the Rubrics of the Missal so as to give a practical norm for the polar regions also. Various hearings were held, practical instructions were given, but the word *aurora* was kept in the rubrics. The Holy See decided that in those regions where there was no dawn the word *aurora* was to be taken in a moral sense, and was to mean that period of time that corresponded to actual dawn. In other words, dawn referred to the beginning of the civil, moral and usual day, the moment when men generally rise and begin their day's work, according to the approved custom of the region.[115]

The decision of the Sacred Congregation of Rites should have been so interpreted, logically, as to apply throughout the year in the polar regions where there never is an actual *aurora*. It was however restricted to the winter or dark months. When the sun shone Mass could be celebrated from midnight.[116] Local time alone was to be used, and determined in the general way, at least during the summer months. The transit of the sun over the local meridian marked midday or noon. To obtain mean time one simply had to reckon with the equation of time.[117] During the winter months, when the sun did not shine, the hours of the day were to be computed in a different way, that is, by the greater brightness of the crepuscules, and especially by the

were reckoned in a much simpler manner, especially before the widespread use of clocks. The day (morning, *mane*) began with the crowing of the cock and finished with that rather vague period between the waning day and the subsequent total darkness when people could no longer be easily and readily recognized.—Cf. Calà, *De Feriis*, q. IV, n. 343.

115 S.R.C., *Missalis Romani*, 18 sept., 2 nov. 1634—*Fontes*, n. 5354. Cf. Ojetti, *Normae Generales*, p. 198, note 4.

116 Cf. Many, *De Missa*, n. 22, 1.

117 *Loc. cit.*

observation of the stars. Once noon had been determined, the reckoning of the other hours was easily made.[118]

Article 7—A Few Additional Remarks

§1—*Tempus Utile*

Very little was said of *tempus utile* before the Code. In general it may be asserted that the Roman Law was accepted with few modifications. Antonellus explained the nature of *tempus utile* as follows:

> Tempus legale duplex est: aliud continuum, aliud utile; continuum dicitur, quod iuxta sonum verbi continue currit diebus feriatis, et non feriatis, in praesentia et in absentia, sive sit copia Iudicis, sive non . . . utile vero dicitur quod [non] currit diebus feriatis, diebus absentiae, et iusti impedimenti, et diebus, quibus non est agendi copia . . . Item aliud est tempus continuum a principio alicuius rei factae, seu faciendae, et in eius progressu, et sic continue continuum; aliud utile a principio, et progressu, et sic continue utile; aliud utile a principio, et continuum in progressu, et sic utile quoad quid, et quoad quid continuum; aliud vero est tempus continuum a principio, et utile in progressu, et sic continuum uno respectu, et alio respectu utile . . . Regulariter vero tempus legale est continuum, praesertim ubi odium consideratur. . . . Quare, ubi lex voluit, quod tempus sit utile, id expressit.[119]

Authors might have been of a different opinion whether for a given institute time was measured as *tempus utile* or *tempus continuum,* but the general teaching has always maintained that time was always presumed to be *continuum* unless the countrary was proved some way or other, as shall be shown in the commentary.[120]

[118] Many, *op. cit.,* n. 22, 2.

[119] *De Tempore Legali,* lib. I, c. II, nn. 1-3.

[120] Some expressions, however, found here and there among the authors are hard to explain in the light of the above mentioned principle. Thus one reads in Antonellus (*op. cit.,* lib. I, c. XV, n. 1): "Verior est opinio, quod tempus tam legale quam conventionale regulariter non currat impedito."—Cf. *Glossa* to *Sciverit* c. 8, *de appellationibus,* II, 15,

§2—The *Dies Certa* and Similar Expressions

The question of the *dies certa* and of other similar expressions is of minor importance. It should therefore be sufficient to quote D'Annibale:

> Porro *dies* in jure plerumque tempus significat: et modo *certa* dicitur, modo *incerta.* Duo videlicet inspicienda sunt, *utrum* et *quando* sit extitura: si neutrum ignoratur, dies dicitur certa, sin alterutrum, incerta. Et si ignoratur, utrum sit extitura, semper pro conditione est. Dies certa, si multiplex potest esse, v.c. *Kalend. Januar., in Paschate,* nec appareat, de qua actum sit, dicitur *indefinita,* et proxima accipitur. Obligationes (ideoque et jura) plerumque ex die initium, aut finem capiunt; et tunc dicitur ex die, vel in diem: *ex die* deberi est post aliquod tempus deberi, adeoque ante illud obligatio non intelligitur: *in diem* vero deberi, est usque ad certum tempus deberi, quo proinde exacto, cessat obligatio. Demum, quoad obligationem praestandam, *diem cedere* dicimus cum aliquid deberi incipit; *diem venire* cum peti potest; scilicet medium tempus relinquitur debitori. Ad ultimum, tempus in jure modo *longum* dicitur, modo *longissimum;* illud decem ann., non plus; hoc 30, non minus spatium continet.[121]

§3—The Expressions *Statim* and *Quamprimum*

The expression *statim* or *incontinenti* takes its place with the most elastic in Canon Law. Antonellus speaks of various opinions according to which the duration of *statim* ranged from one day to one year. The following are explicitly mentioned: one, two, three and ten days; one month and two months; one year.[122] He admits that a few of these opinions are probable, but that it is preferable to leave the matter to the discretion of the judge.[123]

in VI°; S. R. R., *Decisiones Recentiores,* Pars III, 21, n. 10: "Tempus, quando alicui datur, tunc ignoranti illud non currit."

121 *Summula,* I, n. 40.

122 *De Tempore Legali,* lib. I, c. XI, n. 10.

123 "Quamvis autem supradictae opiniones, excepta prima (one month), et ultima (ten days, two months, one year), probabiles sint: probabilior tamen videtur sententia aliorum, id remittentium arbitrio Iudicis, qui ex temporis quantitate, et qualitate facti, et aliis circumstantiis id iudicabit."—*Op. cit.,* lib. I, c. XI, n. 11.

In more recent times the accepted duration of time within which a thing could be regarded as being done *statim* was three days, and the duration indicated by *quamprimum* was determined by the judgment of a prudent man (*bonus vir*).[124]

It might be useful to add here that the reckoning of time in contracts has always been the same according to this principle: "Contracts are recognized as receiving legal sanction from the agreement of the parties." [125]

Article 8—The Immediate Sources of the Code

Before passing on to the study of the Code one might advantageously pause for an instant, and say a few words concerning the immediate sources of the legislation found in Title III of Book I. Canons 31-35 are entirely new in the legislation of the Church inasmuch as general principles for the reckoning of time in Canon Law as a whole had never been previously set down by the legislative authority. Very little, as a matter of fact, was added to what was the common teaching before the codification.

As Van Hove states, the doctrine concerning the duration of time in general as one now finds it in canon 32 was evolved according to Roman Law principles. Likewise does one find in pre-code Canon Law the other Roman institutes of *tempus continuum* and *tempus utile*.[126] The civil and the natural reckonings, as well as the *supputatio prout in calendario*, were used in Canon Law well nigh from the beginning of the Church.

One fundamental change, however, has taken place. The Code clearly states when the natural reckoning is to be used, and when the other is to be followed. Furthermore, rigid rules for the determining of the *terminus a quo* and, implicitly, of the *terminus ad quem*, have relegated to the background the old principle: "Dies coeptus pro completo habetur." These funda-

[124] Cf. D'Annibale, *Summula*, I, n. 167, note 14.

[125] Reg. 85, R.J., in VI°—Cicognani, *Canon Law*, p. 682. The Latin reads: "Contractus ex conventione partium legem accipere dinoscuntur."

[126] *De Temporis Supputatione*, n. 277.

mental rules have been taken, to a great extent, nearly verbatim from the German Code, which one might advantageously quote for future reference:

> Quoad ea, quae de terminis sive legalibus, sive decretis iudicialibus, sive iuridicis negotiis decernuntur valent normae interpretativae, quae sequuntur. (#186)
>
> Si intervallum supputandum initium capiat a facto vel momento, quod non coincidit cum initio diei, hic dies tunc in supputatione negligitur; si vero coincidat, supputatur. Idque etiam valet quoad supputationem diei natalis atque aetatis personarum. (#187)
>
> Intervallum diebus expressum finitur expleto ultimo die; expressum vero hebdomadis, mensibus vel mensium complexu (trimestri, semestri, annali) si quidem eius initium non coincidat cum initio diei, intervallum finitur expleto hebdomadae vel mensis die, qui aut nomine aut numero respondeat diei initiali (tunc non computato); si autem eius initium cum initio diei coincidat, intervallum finitur cum incipit ultimae hebdomadae vel mensis dies, qui aut nomine aut numero diei initiali (tunc computato) respondeat. Quod si mensis die eiusdem numeri careat, tunc semper intervallum finitur expleto ultimo die mensis. (#188)
>
> Nomine "trimestris" vel "semestris" venit intervallum trium vel sex mensium; "dimidiati" vero "mensis" nomine, intervallum quindecim dierum. Si temporis intervallum mense aut mensibus exprimatur cum adjecto dimidiato mense, hic post mensem aut menses integros supputari debet. (#189)
>
> Cum intervallum prorogatur, tempus adiectum expleto praecedente supputandum est. (#190)
>
> Si temporis intervallum, mensibus aut annis constans, non sit continuo supputandum, mensis intelligitur 30 dierum, annus 365. (#191)
>
> Cum loquitur de "initio mensis" dies primus; cum de "mensis die medio," quintus decimus; demum, cum de "fine mensis" ultimus mensis dies intelligitur. (#192)
>
> Si iuridicum negotium perficiendum sit aut praestatio exhibenda die praestituto, aut intra temporis quoddam intervallum, et praestitutus dies aut intervalli dies ultimus dominicam diem incidat aut festum a locali aucto-

ritate recognitum, his diebus sequens profestus dies subrogabitur. (#193)[127]

One thing, however, is not considered by the Code, and that is the problem of when all the various units of time are to be regarded as having lapsed in their entirety or completeness. From the rules set forth in canons 31-35 one knows indeed when and how the various *dies termini* are to be counted, but nothing is mentioned—explicitly at least—about the year and the month. Van Hove makes the statement that, in accordance with the doctrine held before the codification, the Code always requires the various units of time to be completed.[128] It is however to be noted that before the Code the principle that a year once it is begun is then to be held as complete was accepted in many cases.[129] The only way that the Code intervenes in this matter is by using a clearer terminology, thus avoiding ambiguous expressions. An evident instance of a year just begun but held as complete is found in canon 1254, § 2: "Legi ieiunii adstringuntur omnes ab expleto vicesimo primo aetatis anno ad inceptum sexagesimum." [130]

[127] Lib. I, sect. IV—Toso, *Commentaria Minora,* I, 95, note 3. The German Code came into effect in 1900.

[128] "[Codex] semper requirit tempora *completa,* iuxta doctrinam quae apud commentatores recentiores praevaluerat."—*De Temporis Supputatione,* n. 277.

[129] Cf. *supra,* pp. 80-84.

[130] Cf. *infra,* p. 252, where the question is dealt with *ex professo.*

SECTION II

PRACTICAL COMMENTARY OF CANONS 31-35

CHAPTER VI

UNDERLYING PRINCIPLES

ARTICLE 1—INTRODUCTORY REMARKS

All the general norms for the reckoning of time to be found in the Code are condensed in Title III of Book I.[1] The Latin text reads thus:

Can. 31.—Salvis legibus liturgicis, tempus, nisi aliud expresse caveatur, supputetur ad normam canonum qui sequuntur.

Can. 32.—§ 1. Dies constat 24 horis continuo supputandis a media nocte, hebdomada 7 diebus.

§ 2. In iure nomine mensis venit spatium 30, anni vero spatium 365 dierum, nisi mensis et annus dicantur sumendi prout sunt in calendario.

Can. 33.—§ 1. In supputandis horis diei standum est communi loci usui; sed in privata Missae celebratione, in privata horarum canonicarum recitatione, in sacra communione recipienda et in ieiunii vel abstinentiae lege servanda, licet alia sit usualis loci supputatio, potest quis sequi loci tempus aut locale sive verum sive medium, aut legale sive regionale sive aliud extraordinarium.

§ 2. Quod attinet ad tempus urgendi contractuum obligationes, servetur, nisi aliter expressa pactione conventum fuerit, praescriptum iuris civilis in territorio vigentis.

[1] The word *general* is here used in contraposition to *special* and *particular*. General norms apply throughout the Code unless the contrary is expressly provided for, while special norms govern the reckoning of time for this or that institute only. Instances of particular norms are found in canons 1246 and 1635 concerning the reckoning of holy days of obligation and days of fast and abstinence, and for the computation of juridical affairs respectively. Cf. Code's *Analytical Index*, s.v. *Supputatio Temporis.*

Can. 34.—§ 1. Si mensis et annus designentur proprio nomine vel aequivalenter, ex. gr., *mense februario, anno proxime futuro,* sumantur prout sunt in calendario.

§ 2. Si terminus *a quo* nec explicite nec implicite assignetur, ex. gr., *suspensio a Missae celebratione per mensem aut duos annos, tres in anno vacationum menses, etc., tempus* supputetur de momento ad momentum; et si tempus sit continuum, ut in allato primo exemplo, menses et anni sumantur prout sunt in calendario; si intermissum, hebdomada intelligatur 7 dierum, mensis 30, annus 365.

§ 3. Si tempus constet uno vel pluribus mensibus aut annis, una vel pluribus hebdomadibus aut tandem pluribus diebus, et terminus *a quo* explicite vel implicite assignetur:

1°. Menses et anni sumantur prout sunt in calendario;

2°. Si terminus *a quo* coincidat cum initio diei, ex. gr., *duo vacationum menses a die 15 augusti,* primus dies ad explendam numerationem computetur et tempus finiatur incipiente ultimo die eiusdem numeri;

3°. Si terminus *a quo* non coincidat cum initio diei, ex. gr., *decimus quartus aetatis annus, annus novitiatus, octiduum a vacatione sedis episcopalis, decendium ad appellandum,* etc., primus dies ne computetur et tempus finiatur expleto ultimo die eiusdem numeri;

4°. Quod si mensis die eiusdem numeri careat, ex. gr., *unus mensis a die 30 ianuarii,* tunc pro diverso casu tempus finiatur incipiente vel expleto ultimo die mensis;

5°. Si agatur de actibus eiusdem generis statis temporibus renovandis, ex. gr., *triennium ad professionem perpetuam post temporariam, triennium aliudve temporis spatium ad electionem renovandam,* etc., tempus finitur eodem recurrente die quo incepit, sed novus actus per integrum eundem diem poni potest.

Can. 35.—Tempus *utile* illud intelligitur quod pro exercitio aut prosecutione sui iuris ita alicui competit ut ignoranti aut agere non valenti non currat; *continuum,* quod nullam patitur interruptionem.

Authors sometimes speculate on the reasons why these canons on time-reckoning occupy this particular place in the Code. All agree that the first book was the proper place for these canons from the simple fact that they are general norms in accordance with the nature of the first book. Various explanations, however, have been brought forth to show why these canons are found in Title III and not elsewhere in the first book. The most plausible of these is given by Van Hove. He states that no precise reason can be given for this, but that the legislators most likely followed the logical order of inserting this topic where the first application of time-reckoning occurs in the Code.[2]

According to Ojetti[3] and Michiels[4] the probable reason is that the reckoning of time applies to both the *ius obiectivum* and the *ius subiectivum*, and is therefore wedged in between the two, that is, between Titles I and II on the one hand, which treat of the *ius obiectivum*, and the other Titles of the first book on the other hand which consider, above all, the *ius subiectivum*.

Maroto holds a slightly different view. He writes:

> Haec ad solas leges non pertinent, et a Codice traduntur heic in titulo distincto et proprio; sed ea subnecti possent interpretationi legum, quoniam reapse canones de temporis supputatione sunt totidem regulae ad recte interpretandas praescriptiones tempus respicientes in legibus et aliis ecclesiasticis ordinationibus. Ceterum et ea quae dicta fuerunt de interpretatione conveniunt non solum legibus, sed et privilegiis, rescriptis, etc.[5]

[2] These are Van Hove's words: "Ratio autem praecisa cur hoc loco Libri I Codex agat de supputatione temporis, dari non potest. Haec forsan alligari potest: prout in multis aliis materiis, Codex tradit sensum verborum ubi primum in Codice usu veniunt. Ideo Codex titulis de legibus et consuetudine, quae ex se sunt perpetuae, subiungit titulum de temporis supputatione, antequam agat de rescriptis, privilegiis et dispensationibus, quae saepe ad tempus conceduntur."—*De Temporis Supputatione*, p. 238.

[3] *Normae Generales*, pp. 191-192.

[4] *Normae Generales*, II, 125-126.

[5] *Institutiones*, I, 271, note 2.—Cf. Bernareggi, "Metodi e Sistemi delle Antiche Collezioni e del Nuovo Codice di Diritto Canonico,"—*Scuola Cattolica*, Ser. 5, XVIII (1920), 364. This writer rejects Maroto's ex-

In fine it should be said that these explanations are merely of a theoretical nature and should not be given more importance than they deserve. One thing is certain: the norms for the reckoning of time are found in Title III of Book I of the Code. This cannot be changed. It merely behooves one to interpret and apply these canons properly.

Article 2—The Fundamental Norm

The fundamental norm for the reckoning of time is found in canon 31 which reads: "Time is to be reckoned according to the rules of the following canons, unless a different method is expressly provided; the manner of computing time in the liturgical laws remains unchanged."[6] This canon mentions two explicit exceptions to the general rules of time-reckoning found in canons 32-35. Canonists see a third explicit exception in the reckoning of time in contracts.[7] Canon 33, § 2, reads: "The time for determining contractual obligations is to be computed in accordance with the civil law of each country, unless otherwise agreed upon by the contractants."[8]

Implicit exceptions may also be thought of. These occur every time one is confronted with a special form of reckoning not provided for in canons 32-35. A detailed analysis of these four species of exceptions will be given in four subsequent paragraphs. A fifth paragraph will deal with the extension or application in general of canons 32-35 in Canon Law.

§ 1—Liturgical Laws

Liturgical laws, according to Van Hove, are those which ordain

planation on the ground that if it were correct, time-reckoning should be considered in Title I, *de legibus ecclesiasticis.* Cf. Van Hove, *De Temporis Supputatione,* pp. 238-239.

[6] Cicognani, *Canon Law,* p. 672.

[7] Cf. Van Hove, *De Temporis Supputatione,* n. 279, 3, where canon 33, § 2, is associated with canon 31 and is thereby looked upon as a third explicit exception to time-reckoning in general.—The writer, however, would rather consider canon 33, § 2, as affecting merely the reckoning of the hours. Canon 1529 would then be one of the explicit exceptions provided for in canon 31. Cf. *infra,* pp. 117-120.

[8] Cicognani, *Canon Law,* p. 676.

the rites and the ceremonies of the Church.[9] This definition is general and admits of a division which must here be invoked to solve the problem under consideration. There are liturgical laws in the strict sense of the word and those in the broader acceptation of the term. Logically, one could consider liturgical laws in the strict sense those legislative enactments which directly and immediately affect the rites and the ceremonies themselves. Such are the norms found in the Missal and the Ritual for the celebration of Mass and the administration of the Sacraments.[10]

However, such is not the common interpretation of the expression *liturgical laws in the strict sense.* These, according to the common use of the authors, seem to be the ones which are now found only in the liturgical books, such as the Roman Pontifical, the Ceremonial of the Bishops, the Roman Breviary, the Roman Ritual, the *Memoriale Rituum,* and the Roman Martyrology.[11] According to this view the mode of reckoning the liturgical year from the first Sunday in Advent, the determination of the first Sunday of the month for the recitation of the first Nocturn—which, incidentally, sometimes occurs on the last Sunday of the preceding month of the civil calendar—and the enactment of similar norms such as those for the hours according to which especially the choral office is to be recited, would be liturgical laws in the strict sense.

On the other hand liturgical laws in the wide sense would include all the legislative prescriptions found in the Code but affecting the rites and ceremonies of the Church. According to

[9] *Commentarium Lovaniense in Codicem Iuris Canonici,* Volumen I, Tomus II, *De Legibus Ecclesiasticis* (Mechliniae: Dessain, 1930), n. 6.

[10] Liturgical laws in the broad sense according to this classification would refer to those legislative enactments which have only an indirect bearing on the rites and the ceremonies themselves. These laws may be found either in the Code or in the liturgical books. The legal prescriptions which determine the time within which Mass may be celebrated would therefore be liturgical laws in the broad sense.

[11] Cf. Van Hove, *De Legibus Ecclesiasticis,* n. 8, 4. Cf. also canons 2 and 253, § 1.

Maroto such are the application of the *Missa pro populo,* questions of precedence, etc.[12]

Now, it is obvious that the words *salvis legibus liturgicis* of canon 31 apply only to the strict acceptation of liturgical laws. That is made clear by Title III itself. Thus in canon 33, § 1, mention is made of the private celebration of Mass, of the private recitation of the breviary, and of the reception of Holy Communion. These are liturgical prescriptions in the broad sense. One cannot indeed suppose that contradictory statements were contained in the same title of the Code.

Coronata is of the opinion that canon 31 mentions liturgical laws merely *ad abundantiam,* since this matter is treated *ex professo* in canon 2.[13] This is only partly true.[14] The reason for this apparent repetition is that canon 2 in all likelihood stipulates that liturgical laws in general, as contained in the liturgical books, are not considered by the Code and that they continue to retain their binding force unless they are changed by the Code. Canon 31 on the other hand implicitly refers only to liturgical laws in the strict sense. If no mention were made of liturgical laws in canon 31 one might infer from the general tenor of this same canon that the reckoning of time for all liturgical laws without exception was to be made according to canons 32-35, since some liturgical laws are mentioned in canon 33, § 1.

It is therefore to be concluded that only the time-reckoning involved by such liturgical laws as are found in the liturgical books is not affected by Title III.[15] Conversely, the computation of time relative to the liturgical laws found in the Code must be made according to canons 32-35, which have a *vis suppletiva.*

[12] Cf. *Institutiones,* I, n. 122.

[13] *Institutiones Iuris Canonici* (5 vols., Taurini: Marietti, 1928-1936), I, n. 52, 3. This work will hereafter be referred to as *Institutiones.*

[14] Cf. Van Hove, *De Temporis Supputatione,* n. 279, 1.

[15] Such are the determination of Easter, of the first Sunday of Advent, of the first Sunday of the month for the recitation of the first Nocturn. These are all governed by rules found in the rubrics of the breviary. A similar instance is found in the reckoning of the canonical hours such as Prime, Tierce, etc.

Otherwise there would be no norm with which to reckon, for instance, the year that must elapse between the reception of the last of the minor orders and the subdiaconate.[16]

§ 2—Explicit Exceptions

The second exception to the general norms of time-reckoning according to canons 32-35 considers the cases for which a different method is prescribed. Explicit exceptions may be either *ab homine* or *a iure.* The first are provided by the legislator in special cases and are not to be found in the Code. Cicognani mentions such an exception when he writes:

> Thus, Pope Pius XI in his Constitution "*Infinita Dei Misericordia,*" published on the occasion of the Jubilee, decreed that the visits to the Roman Basilicas were to be made either on natural or on ecclesiastical days, that is, commencing with the first vespers of the day and concluding with those of the following day inclusive.[17]

Ab homine exceptions could easily occur in rescripts, privileges and particular precepts emanating from the Holy See.

The *a iure* exceptions are provided by the general law itself when it expressly states that the reckoning of time in a particular case is to be made in a special way and not according to canons 32-35. All the authors undoubtedly agree on the theoretical concept of this exception. It is not so, however, when concrete applications of this norm are considered. The classical illustration of this form of exception is that of the reckoning of time in the gaining of indulgences. Cicognani writes: "Examples of the law providing a method of reckoning time are had in

[16] Cf. canon 978, § 2; Van Hove, *De Temporis Supputatione*, n. 279, 1; Wernz-Vidal, *Normae Generales*, n. 246, note 14.

[17] *Canon Law,* p. 672. The matter here considered is of a liturgical character, but in the broader sense; the reckoning of time for the visitation of churches toward the gaining of indulgences is treated in canon 923 and is normally to be reckoned according to the prescriptions of the Code.

However, this special provision seems to be an explicit derogation to canon 923, and not to canons 32-35.

Canons 921, § 3; 922; 923 and 931 concerning indulgences."[18]

Many authors of note hold a different view.[19] They maintain that the 36 hours which one may use for the visitation of churches in order to gain the indulgences connected with a certain day are not an exception to the general norms of time-reckoning but simply a prorogation of time.

Cicognani states that the reckoning found in canon 923 makes a day consist of 36 hours.[20] This is only partly correct. Canon 923 simply rules that, to gain an indulgence annexed to a certain day, if the visitation of a church or oratory is prescribed, the visit may be made from noon of the preceding day to midnight of the stated day. No mention of a day of 36 hours is made. Thirty-six hours are simply granted for the performance of a definite action. Two weeks might have been given instead of 36 hours, and no one would have considered these two weeks as one day.

As a matter of fact no exception to canons 32-35 exists when nothing more than a certain amount of time is mentioned, such as in canon 923. An explicit exception to the reckoning proper of time as found in canons 32-35 must be made. Such would have been the case if a certain canon of the Code had stipulated that the age required, let us say, to contract a valid marriage, was to be reckoned from moment to moment. That would have

[18] *Canon Law,* pp 672-673. A similar view is held by Maroto, *Institutiones,* I, n. 255; Toso, *Commentaria Minora,* I, 100; Ojetti, *Normae Generales,* p. 194, note 4; Cance, *Le Code de Droit Canonique* (3 vols., Paris: Gabalda, 1927-1929), I, n. 69, c; Wernz-Vidal, *Normae Generales,* n. 246, note 15; Cocchi, *Commentarium in Codicem Iuris Canonici ad Usum Scholarum,* Vol. I (2. ed., Taurinorum Augustae: Marietti, 1921), n. 93. This work will hereafter be cited as *Commentarium.*

[19] Cf. Vermeersch-Creusen, *Epitome,* I, n. 148; Cappello, *Summa Iuris Canonici in Usum Scholarum Concinnata,* Vol. I (2. ed., Romae: apud Aedes Universitatis Gregorianae, 1932), n. 179, 2, 4. This work will hereafter be referred to as *Summa.* Cf. also Van Hove, *De Temporis Supputatione,* n. 279, 2, who justly remarks that the matter is without importance.—The other canons dealing with indulgences hardly deserve mention here from the viewpoint now considered.

[20] *Canon Law,* p. 673.

been an explicit exception to canon 34, § 3, according to which one's age is always reckoned from day to day.

Clear cut *a iure* exceptions are not easily found in the Code. Provision for a reckoning other than that which is prescribed in canons 32-35 is made in canons 1508 and 1520, where the civil law is canonized relative to prescription and contracts respectively. The special exception for contracts is to be treated in the following paragraph.

§ 3—Contracts

The universal teaching of the canonists admits that the reckoning of time in contracts is in some way or other exceptional. That is provided for in canon 1529 which reads:

Quae ius civile in territorio statuit de contractibus tam in genere, quam in specie, sive nominatis sive innominatis, et de solutionibus, eadem iure canonico in materia ecclesiastica iisdem cum effectibus serventur, nisi iuri divino contraria sint aut aliud iure canonico caveatur.

In general, therefore, civil law is to be followed for the reckoning of time in contractual matters. As a rule the civil laws give the contracting parties full liberty in determining not only the duration of time for these obligations, but also its reckoning. That is a universally accepted legal principle based on the natural law itself.[21]

These principles are admitted in American civil law: "The time for performance of a contract is governed by its terms."[22] "It is ordinarily regarded as legal for the parties to stipulate as

[21] "Contracts are recognized as receiving legal sanction from the agreement of the parties."—Reg. 85, R. J. in VI°—Cicognani, *Canon Law,* p. 682. Cf. Ojetti, *Normae Generales,* p. 200, 8; Michiels, *Normae Generales,* II, 134.

[22] *Corpus Juris Secundum, A Complete Restatement of the Entire American Law as developed by All Reported Cases* (19 vols., Brooklyn: The American Law Book Co., 1936-), Vol. XVII, Contracts #502. This work will hereafter be cited in the abbreviated form: 17 C.J.S. Contracts #502.

to the time when actions may be brought on their contracts." [23]

However, this liberty granted to the contractants must comply with statutory prescriptions which stipulate that certain legal formalities are to be followed in certain instances. Thus: "Notwithstanding an oral agreement as to the terms, there is no binding contract where law requires a contract under seal." [24]

Then, also, the civil law concept of month and day might have a special meaning. That legal interpretation is to be adhered to unless the contrary is explicitly mentioned. Thus, in English law, a month in a deed is 28 days, unless the context shows that a calendar month of 31 days was intended.[25]

Augustine writes; "Thus also, according to English law, when a calendar month's notice of action is required, the day on which it is served is included and reckoned one of the days; and therefore, if a notice be served on the 28th of April, it expires on the 27th of May, and the action may be commenced on the 28th." [26]

If there were no other canonical enactment concerning contracts than that of canon 1529, the question would well nigh be without complication. However, canon 33, § 2, rules that: "The time for determining contractual obligations is to be computed in accordance with the civil law of each country, unless otherwise agreed upon by the contractants." [27]

Now, one may doubt whether this regulation, which gives the parties the right to use a time-reckoning different from that

[23] 17 C.J.S. Contracts #531. Cf. also #358, #385, #396, #397, #398, #399, #402, #431, #503, #504, #632.

[24] 17 C.J.S. Contracts #63. Contracts under seal are generally required in bonds, deeds, etc. Cf. 17 C.J.S. Contracts #242: "An agreement unlimited as to both time and space is invalid as a total restraint of trade."

[25] Cf. Augustine, *A Commentary on the New Code of Canon Law* (8 vols., Vol. I, 3. ed., St. Louis: Herder, 1920), p. 118. This work will hereafter be cited as *Commentary*.

[26] *Commentary,* I, 118. Cf. Toso, *Commentaria Minora,* I, 106-108, where the laws of various countries are given concerning contracts. What has been said thus far for contracts applies *servatis servandis* to prescription also, according to canon 1508.

[27] Cicognani, *Canon Law,* p. 676.

prescribed by the civil law, modifies exclusively canon 33, § 1, or whether it should apply to the reckoning of time in general. Many authors admit the latter, implicitly at least.[28] Van Hove is explicit on this point. He states clearly that the norm contained in canon 33, § 2, is general, and is therefore to be applied not only to the reckoning of the hours of the day, mention of which is made in canon 33, § 1, but to the reckoning of time in general.[29]

In practice this question is of little importance, since the duration and reckoning of time in general nearly always depend on the will of the contractants. However, the position of this regulation seems to indicate that this particular exception refers only to the reckoning of the hours of the day, since it is placed in a canon that deals exclusively with the computation of these units. If this prescription had been intended to affect the reckoning of time in general, it would have been placed within, or immediately after, canon 31, which lays down the general norms for the computation of time. The same conclusion is to be drawn from the wording itself of canon 33, § 2. Thus, the Code uses the singular—*praescriptum iuris civilis*. This would normally indicate that one particular point only, and not the reckoning of time in general, was involved.

The special exception contained in canon 33, § 2, is more easily understood if it applies only to the reckoning of the hours. The only liberty that the Code grants to private individuals in choosing a time-reckoning that differs from that established by common usage is in the computation of the hours of the day. The weeks, months and years are of themselves necessarily reckoned according to the computation stipulated by law. It is not surprising then that, in general, contractants are free to choose a time-reckoning of their liking only insofar as the civil law permits it.[30] However, when the reckoning of the

[28] Cf. Michiels, *Normae Generales,* II, 134; Chelodi-Bertagnolli, *Ius de Personis* (2. ed., Tridenti: Ardesi, 1926), n. 89; Coronata, *Institutiones,* I, n. 52, 2, a, etc.

[29] *De Temporis Supputatione,* n. 279, 3.

[30] Cf. canon 1529.

hours is involved, it would seem that a special computation could be chosen even if the civil law forbade it.

The provision contained in canon 33, § 2, seems to have been purposely inserted to prevent the possibility of needless litigation when the contractants have made no mention of a special reckoning. In that case the computation that the civil law prescribes in such instances is to be the accepted one in ecclesiastical tribunals.

Another merely theoretical question closely connected with the preceding one is that of the binding force of the words "*nisi aliter expressa pactione conventum fuerit.*" Thus, if the law of a certain country did not allow the contractants to choose a time-reckoning different from the legal one, but the parties nevertheless did so, which would be the acceptable computation before an ecclesiastical court: the legal reckoning prescribed by the civil law, or that chosen by the contractants?

Van Hove maintains that the reckoning prescribed by the State would be the one to follow.[31] The more common opinion of the canonists, however, holds that canon 33, § 2, gives the contractants' selection priority over the prescriptions of the civil law, if such a prescription exists in the law.[32] That seems to be a logical conclusion. Indeed, the civil laws of a State bind in ecclesiastical courts only insofar as they are canonized. Canon 1529 thus canonizes the prescriptions of civil law in contractual matters, but does not do so in an absolute manner. There is a restriction in the words "*nisi. . . aliud iure canonico caveatur*". There is no reason to believe that canon 33, § 2, is not one of these exceptions.

[31] "Si tamen in certis regionibus hae dispositiones [civiles] sint cogentes. et proinde paciscentes his derogare non possint, opinemur canonem 33, § 2 disponere tantum de casibus communiter contingentibus, in quibus derogatio privatis permittitur et non derogare principio generali c. 1529, seu principium non esse hic invocandum: generi per speciem derogatur."—*De Temporis Supputatione*, n. 279, 3.

[32] Cf. Michiels, *Normae Generales*, II, 134; Ojetti, *Normae Generales*, p. 200, 8; Vermeersch-Creusen, *Epitome*, I, n. 144; Maroto, *Institutiones*, I, n. 255.

One could therefore conclude that the prescriptions of the civil law relative to the reckoning of time in contracts actually bind, with the exception of those laws that prevent one from choosing any one of the computations found in canon 33, § 1. Thus, if the civil law prescribes certain formalities or a certain age for contracts, these regulations are to hold in ecclesiastical courts also. Conversely, if the use of Standard Time, let us say, is necessarily prescribed for the reckoning of the hours of the day during which a contract is to be performed, these do not bind and the parties are free to choose local time or any other mentioned in canon 33, § 1.[33]

Some authors, to solve the apparent contradiction between canons 33, § 2, and 1529, give to the words *"praescriptum iuris civilis"* of canon 33, § 2, a meaning that differs somewhat from the customary acceptation of these terms in Canon Law. The expression now under examination would refer merely to private law prescriptions which are found, for instance, in commercial codes, and not to the civil law proper of the State.[34]

A final interpretative difficulty in this matter is concerned with the meaning of the words *urgendi contractuum obligationes.*[35]

[33] In American civil law full liberty is given the contractants relative to the determination of time. In certain instances, however, if no provision is made to the contrary, the day is said to end with the setting of the sun. Cf. 17 C.J.S. Contracts #482: "To make a tender good as to time of day, the general rule is that the tenderer must, at the latest time, on the last day of the term of the contract, before the sun sets, produce the money or goods and offer to comply with the contract, and the tender must be made a sufficient length of time before the sun sets so that the money may be counted or the goods examined by daylight."

[34] Cf. Toso, *Commentaria Minora,* I, 105; Cicognani, *Canon Law,* p. 682. —Van Hove (*De Temporis Supputatione,* n. 279, 3) rejects such explanations when he writes: "Ad solvendam hanc difficultatem, A. Toso [*Commentaria Minora,* I, 105] praescriptum iuris civilis intelligit non legem Status civilis, sed dispositiones *iuris privati* tantum legum Civitatum, quae in Codicibus iuris civilis vel commercialis continentur, quibus per pactiones privatorum derogari potest. Sed hic sensus vocabuli 'civilis' est alienus omnino a Codice iuris canonici. Hic legem civilem semper intelligit legem Status civilis."

[35] Cf. Van Hove, *De Temporis Supputatione,* n. 279, 3; Eppler, *Quelle*

A few authors maintain that these words are to be taken in a wide sense so as to include all contractual obligations according to canon 1529. One would then have the faculty of choosing any acceptable reckoning, not only with reference to the time of bringing an action, but also with reference to the time within which the contract is to be performed.[36]

Other canonists, however, are of the opinion that the word *urgere* is to be taken in a restrictive sense so as to mean the exacting of the fulfilling of contractual obligations. *Urgere* would then mean *exigere*.[37]

This restrictive meaning of *urgere* is found in canon 1686, where it is ruled that the victim of fear or deceit in a contract is granted the *exceptio metus vel doli* if and when the execution of a contract is urged. Similar instances of this particular acceptation of *urgere* are found in canons 24, 336, § 1, 2251, 2347.

There does not seem to be any reason for restricting the meaning of *urgere* in canon 33, § 2. This concession of choosing any one of the prescribed reckonings is a favor and should not be limited by private individuals unless there were a canonical justification for so doing.

§ 4—Implicit Exceptions

These exceptions are not explicit, but may be deduced from

und Fassung des katholischen Kirchenrechts mit einem Anhang ueber seinem zeitlichen Geltungsbereich (Zürich: 1928), p. 53, note 160.

[36] Wernz-Vidal, *Normae Generales*, n. 246, note 16; Hilling, *Die allgemeinen Normen des Codex Iuris Canonici* (Freiburg i. Br.: Verlag von Josef Waibel, 1926), p. 153; Falco, *Introduzione allo Studio del "Codex Iuris Canonici"* (Torino: Fratelli Bocca, 1925), p. 91—Van Hove (*De Temporis Supputatione*, n. 279, 3, note 4) also mentions as holding this view Eichmann (*Lehrbuch des Kirchenrechts auf Grund des Codex Iuris Canonici* [3. ed.], p. 48).

[37] Cf. Toso, *Commentaria Minora*, I, 104; Coronata, *Institutiones*, I, n. 52; Blat, *Commentarium Textus Codicis Iuris Canonici* (6 vols., Romae, 1921-27), I, 120. This work will hereafter be cited as *Commentarium*.—Van Hove (*De Temporis Supputatione*, n. 279, 3, note 5) also mentions Cicognani, *Ius Canonicum*, II, 191. He adds that many commentators simply mention the words of the canon without commenting on them.

the general tenor of the Code. Thus, a properly established custom might reckon from moment to moment the age of those who are to receive the sacred orders.[38]

Other implicit exceptions might be thought of. These, however, would not be absolute, but merely relative. Their existence depends on the concept that one has of canons 31-35 as a whole. Thus, if one maintains that an absolute and physical completion of time-intervals is prescribed in canons 31-35,[39] one must necessarily admit that canons 766, n. 1, and 1254, § 2, are exceptional, since it is sufficient to reach, and not necessary to complete, the last year of the time-interval mentioned by the law.[40]

Likewise, if one admits that the physical reckoning is the rule in the Code as imposed by Title III, one must of necessity look upon canon 788 as an exception.[41]

There remains another problem. A careful study of canons 31-35 leaves one under the impression that the framers of the Code, when Title III was written, had in mind only the classical cases in which the reckoning of time is made according to the civil or juridical calendar.[42] Such would be the computation followed in the reckoning of one's age, in the reckoning of a period of time expressed in days, weeks, months, or even in years, in those instances in which this time-unit is not taken in a special sense.

Canon 31, as has been already shown, is very general, and rules that the reckoning of time is to be made according to

[38] Cf. canons 25-30.

[39] Cf. Lacau, *De Tempore,* p. 34. Cf. *infra,* pp. 252-254.

[40] Cf. Van Hove, *De Temporis Supputatione,* n. 280, where canons 766, n. 1, and 1254, § 2, are looked upon as express exceptions to the general norms of time-reckoning.—The matter is undoubtedly of no practical importance, but the present writer takes a different view. He believes that the above mentioned canons are in no way contrary to the general prescriptions of Title III. Cf. *infra,* p. 252.

[41] Much remains to be said about the completion of a specified interval of time and the problem of the physical computation, but the writer has found it expedient to treat of them in Chapter VIII, Article 2, once the explicit interpretation of canons 31-35 has been given.

[42] Cf. canon 32, § 2.

the prescriptions contained in canons 32-35. Three explicit exceptions to this law, however, are expressly mentioned, namely, liturgical affairs (strict sense), express legal provisions to the contrary, and contracts.

Now, one finds in the Code certain time-intervals— consisting of a year especially—which are not reckoned according to canons 32-35, and which, at the same time, cannot be considered exceptional according to canons 31 and 33, § 2.

If the time-intervals mentioned in canons 1365, 976, § 2, 978, § 2, are to be reckoned according to the prescriptions of Title III, one must then give the word *calendar* a somewhat elastic meaning which cannot be conceived as having entered the mind of the codifiers. The legal term *calendar* has indeed always referred to the civil calendar only.[43]

§ 5—Extension of the General Norms

The rules contained in canons 32-35 are to be followed throughout the Code except in the cases singled out above. Furthermore, the same norms apply to all acts emanating from the Holy See, such as privileges, dispensations, rescripts, precepts, decrees, special laws and the like.[44]

These laws, however, do not bind *per se* the inferior legislators who are acting within the limits of their jurisdictional authority. Duration and reckoning of time go hand in hand, and when one is free to grant such or such a period of time for vacations, let us suppose, then one is also free to determine the reckoning that is to be used in the computation of the time-interval granted.[45]

A bishop, for instance, could grant a vacation of fifteen days to be reckoned from 6 P. M. July 1 to 6 P.M. July 16. In this case the starting point would otherwise be explicitly determined, and this time-interval would normally follow the civil and not the natural computation of the day. The vacation would customarily end at midnight July 16.

[43] Cf. *infra,* pp. 191-199.

[44] Cf. Maroto, *Institutiones,* I, 271, noet 2; Michiels, *Normae Generales,* II, 132-133.

[45] Cf. Michiels, *op. cit.,* p. 133.

The same remarks also apply to judges. In many instances the judge is free to determine the duration of a penalty or the duration of the interval that is conceded to the parties to perform certain acts. Such is the case in canon 2321:

Sacerdotes qui contra praescripta can. 806, § 1, 808 praesumpserint Missam eodem die iterare vel eam celebrare non ieiuni, suspendantur a Missae celebratione ad tempus ab Ordinario secundam diversa rerum adiuncta praefiniendum.

The ordinary is here free not only to determine the amount of time for the suspension but also the mode of its reckoning.[46]

In other cases the precise amount of time is determined by the Code. An example of this is found in canon 2410, according to which religious superiors are *ipso facto* suspended for a month from the celebration of Mass if and when they violate the prescriptions of canons 965-967, by knowingly having their subjects ordained by another ordinary than the local ordinary. The judge would merely give a declaratory sentence in determining whether or not the penalty was actually incurred. If it is, its duration and reckoning depend on the Code. Similarly, the judge is not free to grant an extension to or to provide with a special reckoning the time-periods mentioned in canons 1736, 1764, § 4, 1847, 1893, 1905.[47]

When the law prescribes a certain period of time but gives the judge the faculty of prolonging it in certain instances, the extension granted partakes of the nature of the original period and must necessarily be reckoned in accordance with it, that is, according to the general norms of the Code.[48]

[46] Cf. Toso, *Commentaria Minora,* I, 102: "Tempus *concessum ab homine, diversum reputatur ab eo, quod a iure statuitur, et ex uno ad aliud non infertur* (S. Rota, in *Recentior.* decis. 13 n. 8 part. 13)."

[47] Cf. Roberti, *De Processibus* (2 vols., Romae: apud Aedes Facultatis Iuridicae ad S. Apollinaris, 1926), I, n. 180.

[48] Cf. Toso (*Commentaria Minora,* I, 102): "*Contra, tempus prorogatum* (idest ab homine), *censetur unum et idem cum primo tempore* (*ibid.* [S.R. Rota, *Decisiones Recentiores*] decis. 74 n. 26 part. 16 et alibi), *et omnibus pollet qualitatibus ac iuribus, perinde ac si idem terminus substantifice fuisset* (S. Rota, in *Nuperr.* decis. 289, n. 11).—Such extensions of the

Article 3—The Duration of Time

Canon 32 deals with the theoretical length of the various time-units.

§ 1—The Day

This time-unit consists of 24 hours to be reckoned without intermission from midnight to midnight. If these words are accepted as they stand without proper interpretation, one cannot find any possibility of reckoning time from moment to moment, as is stipulated in canon 34, § 2. Thus, if each and every juridical day is to be reckoned from midnight to midnight without exception, it becomes evident that no day could be reckoned from 10 A.M. to 10 A.M. or from any other hour of the 24.

To solve the apparent contradiction between canons 32, § 2, and 34, § 2, many authors state that the codifiers did not intend, in canon 32, § 2, to determine the reckoning proper of the day, but simply to mention its duration.[49] The present writer, however, does not quite agree with this view, because even if one admits that canon 32, § 2, considers only the duration of time, all the difficulties are not solved. The duration, for instance, of the true solar day is not exactly of 24 hours.[50]

original legal time-interval may be had in view of canons 1883, 1676, § 3, 1922, § 3. Cf. Roberti, *De Processibus,* I, n. 180.

[49] Cf. Van Hove, *De Temporis Supputatione,* n. 282; Michiels, *Normae Generales,* II, 150.—Vermeersch held for a number of years that the day was always to be reckoned from midnight to midnight as an indivisible unit. This theory is expounded especially in the article: "De Nonnullis Supputationibus ad Usum Religiosorum,"—*Periodica,* XVII (1928), 80*-82*. The same opinion was also expounded in the earlier editions of the *Epitome.* Cf. *Epitome* (4. ed.), I, n. 116. In the fifth edition, however, that doctrine is abandoned. Cf. I, n. 146.

It is admitted by every one today that the day may be reckoned from moment to moment Cf. Toso, *Commentaria Minora,* I, 109; Ojetti, *Normae Generales,* pp. 193-194; Coronata, *Institutiones,* I, n. 48; Michiels, *Normae Generales,* II, 149-150; Van Hove, *De Temporis Suppuatione,* n. 282.—Beste (*Introductio in Codicem* [Collegeville, Minn.: St. John's Abbey press, 1938], p. 100) holds that even the hours are reckoned according to either one of the computations.

[50] Cf. *infra*, pp. 127-128.

Canon 32 sets down for the day certain restrictions that are not applied to the other units. Thus, the 24 hours of the day are to be reckoned without interruption. The reason for this is most likely that the day in Canon Law is considered an indivisible unit, the ultimate sub-division of time. One could infer that a vacation of one month or two weeks, unless otherwise specifically determined, could be interrupted. Five or six days could be taken at one time, and the remainder of the vacation later on. Conversely, these vacation days cannot be broken up. Thus, a pastor is not free to be away from his parish for eight hours on three different occasions and maintain that he has taken only one day's vacation.

If one were to admit that the day is not an indivisible unit strange conclusions would follow. Thus in canon 338, § 2, one reads that bishops have a right to take a vacation of not more than three months a year, and that these months are either continuous or intermittent. In other words, bishops are not obliged to take the whole of the three months at the same time, but may divide this period into two, three, four or more sections according to their good pleasure, provided their absence is not detrimental to their diocese. That does not mean, however, that the bishops could be absent from their dioceses 90 x 24 or 2160 hours, as could be logically upheld if one admitted that the day is not an indivisible unit. If that were true, a bishop could be away from his diocese six hours a day all year. This is obviously contrary to reason.

In this matter, therefore, a day even partially spent is held as the equivalent of a totally elapsed day. That is precisely what the Congregation of the Council decreed with reference to the vacations of canons. The following doubts were proposed:

> II—Since it was decided in *Toletana et Aliarum*, 20 [10] July, 1920, that the time of absence should be computed by whole days, the question now is whether the days on which a canon is illegitimately absent for some of the hours, but not for all of them, is to be counted as a day of residence or a day of absence.
>
> III—On those days on which a canon is illegitimately absent for some of the hours, does he lose:

a) only the distributions corresponding to the hours when he is absent; or

b) the fruits and distributions in proportion to the hours of illegitimate absence; or finally,

c) the fruits of the whole day and the distributions for the hours when he was illegitimately absent.

Reply II. In the negative to the first part; in the affirmative to the second part.

III. In the negative to the first and second parts; in the affirmative to the third part.[51]

It is made clear that the canon who is illegitimately absent from a single canonical hour loses the fruits (not necessarily all the distributions) for the whole day.[52]

One may logically conclude that the same rule applies, *servatis servandis,* to the reckoning of the day in general, for instance, to the absence of bishops, pastors, etc. It is true that canon 32, § 1, mentions the hours of the day, but it is to be noted that the duration of the hour is not determined as in the case of the day, week, month and year. The hour is here considered merely as constituent part of the day, which is *de facto* the smallest indivisible unit.[53]

This interpretation is consonant with the text of canon 32, § 1. In the description of the day one finds three distinct elements:

a)—a duration of 24 hours;

b)—a continuous reckoning;

c)—a reckoning that extends from midnight to midnight.

Now, if any canon contains an implicit derogation to any one of these elements, it does not necessarily destroy the three

[51] S. C. Conc., *Resolutio, Abulen. et aliarum,* 16 mart., 1924—*AAS,* XVII (1925), 192—Bouscaren, *Canon Law Digest,* Vol. I (Milwaukee: Bruce, 1934), pp. 233-234. Cf. also S.C. Conc., *Toletana et aliarum. Servitii choralis,* 10 iulii 1920—*AAS,* XII (1920), 357-365; S. C. Conc., *De Zacatecas et aliarum. Residentiae et distributionis,* 16 martii 1924—*Monitore Ecclesiastico,* XL (1928), 289-297.—Bouscaren's work will hereafter be cited by the letters *CLD.*

[52] Cf. Van Hove, *De Temporis Supputatione,* p. 274, note 2.

[53] Cf. Coronata, *Institutiones,* I, n. 50.

of them. Thus, in canon 34, § 2, provision is made for the moment to moment reckoning. Consequently, in such instances the day is to be reckoned from any one of the hours of the day. Element c does not here apply, but elements a and b remain in force, and the day still consists of 24 continuous hours.

Likewise, in canon 33, § 1, an implicit exception is made to the duration of the 24 hours, since the true solar day does not consist precisely of 24 hours. During a good part of the year a few seconds would be lacking.[54] When a community or State shifts from Standard to Daylight Saving Time, one hour is dropped from the 24, and thus such a day according to the usual computation has not more than 23 hours.

Augustine seems to admit as a general principle that the Code considers fractions of a day when he writes: "Canon Law, by enjoining computation from moment to moment, if nothing is said to the contrary, considers fractions." [55] Michiels states explicitly that the days of one's vacation may be reckoned intermittently. Thus if one day's vacation is interrupted, the remaining hours may be taken on some other occasion.[56]

Canonical equity might in some instances allow one to follow Michiels' interpretation, but the letter of the law is certainly opposed to it. A derogation to one of the three elements of the day does not affect the others.

In this matter it seems that the old principle *"parum pro nihilo reputatur"* is maintained, at least in certain cases. Thus, if a pastor commenced his vacation and was, for some reason or other, forced to return after two or three hours, that time would most likely not count.[57] On the other hand, when one is absent from morning till evening, then a full day must certainly be deducted. The reckoning of vacation days is somewhat analogous

[54] Cf. *supra*, pp. 41-46.

[55] *Commentary*, I, 122.

[56] "Etsi de die nihil expresse statuatur, videtur tamen omnino affirmandum, quod compleri possunt 24 horae de facto interruptae."—*Normae Generales*, II, 154.

[57] Cf. Claeys-Boúúaert, *Selecta Capita Codicis Iuris Canonici* (Gandae, 1919), p. 56. This work will hereafter be referred to as *Selecta Capita*.

to the reckoning of available time (*tempus utile*). A notable part of the day means a full day.[58]

Before passing on to the other time-units the writer would like to add a word here about canon 1246, which rules that the reckoning of holy days of obligation and of days of fast and abstinence is from midnight to midnight. At first sight it would seem that this is a tautology and merely an application of canon 32, § 1, and that nothing would have been changed if no special provision had been made for these institutes. A further analysis, however, reveals that this is not the case. It has been shown in Section I that the only existing laws concerning days of fast and abstinence stipulated a reckoning from sunset to sunset, but that custom gradually brought about the use of the midnight to midnight computation. The Code therefore wisely intervened in this matter to remove all doubts and to establish once and for all that these days were to be reckoned everywhere according to the general norms of canons 31-35.

§ 2—The Week

This time-unit hardly deserves mention. It is the only one that does not cause trouble. Canon 32, § 1, simply states that the week consists of 7 days. There is no determinate starting point as for the day, and the days are not necessarily reckoned continuously. In other words, the week is not necessarily reckoned from Sunday to Sunday, as is done in the liturgy, and the seven days may be intermittent if a continuous reckoning is not otherwise prescribed.[59]

This is clearly shown by the fact that the words *"continuo supputandis"* are not applied to the week. This time-unit is reckoned in the same way as the month and the year, except that there can be no question here of the computation according to

[58] Cf. *infra*, pp. 237-239.

[59] Cf. Van Hove, *De Temporis Supputatione*, n. 283; Maroto, *Institutiones*, I, n. 257, 7; Michiels, *Normae Generales*, II, 150; Vermeersch-Creusen, *Epitome*, I, n. 147.

the calendar in the strict sense, since the week invariably consists of 7 days.[60]

§ 3—The Month and the Year

Canon 32, § 2, states that a month and a year at law consist of 30 and 365 days respectively unless the law prescribe that these units be computed according to the calendar. In that case one finds that the month has either 28, 29, 30 or 31 days, while the year has either 365 or 366 when the Gregorian calendar is used. It might be noted here that canon 32, § 2, seems to imply that the juridical month and year (30 and 365 days) are the rule in Canon Law and that the reckoning according to the calendar is exceptional. In practice, however, as shall be shown subsequently, the computation according to the calendar is certainly the more often used.

Coronata remarks pertinently that 12 so called juridical months of 30 days do not make up one juridical year of 365 days.[61]

[60] In certain instances, however, if a determinate week were prescribed, the time would then be reckoned from Sunday to Sunday, as in the calendar. Vidal seems to be the only author to imply that the week must in every case be reckoned without intermission. Cf. Wernz-Vidal, *Normae Generales,* n. 247.

[61] *Institutiones,* I, n. 49, 1. Authors often bring up another question at this point, namely, that of the leap day. It has been found advantageous to consider this problem when the calendar is treated *ex professo.*

CHAPTER VII

THE RECKONING OF THE HOURS

This chapter deals with a practical question for many of the faithful. While the reckoning of the other units pertains to canonists above all, the computation of the hours confronts the private individual on many an occasion. Whether it be for that reason or any other, the fact remains that this topic has an abundant literature, at least relatively to the other questions considered in Title III.

Canon 33, § 1, rules that:

> In reckoning the hours of the day, the common custom of the place is to be followed; but in the private celebration of Holy Mass, in the private recitation of the Divine Office, in receiving Holy Communion and in the observance of the laws of fasting and abstinence, one may deviate from the common custom of the place and follow the local time, true or mean, or the legal time, regional or extraordinary.[1]

The general tenor of this paragraph is easily understood. Only the details offer a difficulty of interpretation.

ARTICLE 1—THE INSTITUTES INVOLVED

The nature of the four institutes explicitly mentioned shall now be considered.

§ 1—The Private Celebration of Mass

The Code determines the time-interval within which the Holy Sacrifice of the Mass may be offered. In ordinary circumstances Mass is not to begin earlier than one hour before dawn, nor later than one hour after noon.[2]

[1] Cicognani, *Canon Law,* p. 676.

[2] Canon 821, § 1. A through explanation of the meaning of dawn is given *supra,* pp. 101-103, with the rules to be followed when dawn does not actually occur.

The general law provides an exception to this rule. According to canon 821, § 2, on Christmas day conventual or parochial Masses may be begun at midnight, but no other Mass without an Apostolic indult. However, § 3 rules that in all religious or pious houses having an oratory with the faculty of reserving the Blessed Sacrament there habitually, one priest may say at midnight on Christmas three Masses according to the liturgy of the day, or only one; this Mass satisfies the obligation of hearing Mass for all who assist at it, and Holy Communion may be given to those who wish to receive it.[3]

There is no difficulty here concerning the beginning of dawn. This is a natural phenomenon which is not susceptible of various reckonings such as are found in canon 33, § 1. Dawn is the pivot upon which hinges the determination of the time before which Mass may not be celebrated in ordinary circumstances.[4] It is the same for both the private and the public celebration of Mass. The distinction between the private and the public celebration can be invoked only in the reckoning of midnight and, above all, in the computation of noon.

Unfortunately, the Code does not throw any light on the meaning of *private* and *public* with reference to the celebration of

[3] Cicognani (*Canon Law*, pp. 677-678) mentions special laws concerning the time when Mass may be begun. He writes: "Besides this concession granted by the Code, with reference to midnight Mass on extraordinary occasions, the Holy See has given general permission on the occasion of Eucharistic Congresses, if the Blessed Sacrament remains exposed for public veneration throughout the whole night, that one Mass be said at midnight and at it all may receive Holy Communion. The priest who takes part in the nocturnal adoration may say Mass immediately after the one Mass celebrated for the people, or one hour after midnight. As to midnight Mass on other occasions, the Sacred Congregation of the Sacraments can grant permission under the following conditions: (1) on extraordinary occasions only; (2) the Mass may not begin until half an hour after midnight; (3) the adoration at night must be continued for about three hours; (4) all danger of irreverence must be removed."—S.C. de Sacramentis, 22 aprilis 1924—*AAS*, XVII (1925), 100.

[4] Cf. canon 821, § 1. Mass may not be begun earlier than one hour before dawn.

Mass. No wonder, then, that various interpretations have been expounded.

According to some authorities private is taken in a liturgical sense and qualifies any Mass that is not sung.[5] Others maintain a slightly different view and state that a private Mass is one that is not a solemn Mass, that is, a Mass sung with the assistance of deacon and subdeacon as ministers at Mass.[6]

Another school teaches that private is taken in a juridical sense, and therefore applies to any Mass that is not prescribed by ecclesiastical law,[7] or that is not annexed as a duty to an office by ecclesiastical law.[8] Public Masses would then be capitular, conventual or parochial Masses.[9]

Michiels agrees substantially with this, but gives a somewhat different explanation. He states that a private Mass, insofar as it is opposed to a public one, is any Mass, whether read or sung, or even solemn, which is not, in virtue of some ecclesiastical law or precept, annexed to an office which must *per se* be exercised publicly.[10]

Van Hove maintains a similar opinion and rejects the liturgical acceptation of "private" because the celebration of a high or solemn Mass is not of its own nature necessarily ordained for the public use of ecclesiastical communities, but may also be intended for private devotion.[11] He seems to found his opinion on

[5] Thus Toso, *Commentaria Minora*, I. 103; Blat, *Commentarium*, I, n. 95.

[6] Thus Cicognani, *Canon Law*, p. 677; Ojetti, *Normae Generales*, p. 199.

[7] Thus Cappello, *Summa*, I, n. 179, 2, 1.

[8] Thus Vermeersch-Creusen, *Epitome*, I, n. 148; Cardenas, "Midnight and Canon Law,"—*ER*, CI (1939), 402, n. 1.

[9] Cf. Beste, *Introductio*, p. 100.

[10] *Normae Generales*, II, 137. Cf. Vromant, *Ius Missionariorum. Introductio et Normae Generales* (Lovanii: Museum Lessianum, 1934), n. 133, 2, note 2. This work will hereafter be cited as *Normae Generales*.

[11] *De Temporis Supputatione*, n. 290; "Proinde iuxta usualem supputationem celebrandae sunt Missa capitularis (can. 413, § 2 et 3), conventualis, (can. 610, § 2) et paroecialis, aut etiam quaevis Missa quae, hora statuta per legem universalem vel particularem, est celebranda ad usum publicum fidelium etiam si in cantu non celebretur. Huiusmodi est Missa

the assumption that when a Mass is to be celebrated in virtue of one's office (*vi officii*) that Mass is, of its own nature, ordained for the public use of the community. Consequently, he adds, if the public is to assist at that Mass, the usual reckoning of the hours is to be followed. This argumentation is not conclusive. In the United States there is an actual proof to the contrary. It is a well known fact that nearly all the American trains run on Standard Time throughout the year, even in those States or cities where Daylight Saving Time is the usual one during the summer months. No one seems to be in the least perturbed by this state of affairs.

Vidal strikes a different note. He considers the word private in the common every-day meaning and holds that a private Mass is one that is celebrated merely through devotion.[12] From the above norm this author does not draw any other conclusion than stating that capitular, conventual and parochial Masses are necessarily public ones.

There is much to commend this acceptation of the word *private.* It seems to be in conformity with canon 18.[13]

The writer would conclude that any Mass to which the community or the public in general is invited, is not considered a private Mass. Thus all the Masses, low or otherwise, that are scheduled for a certain hour, even on week days, would be looked upon as public Masses. The priest who has agreed to celebrate Mass at 8 o'clock could not celebrate at a different hour without a sufficient reason. He has a moral obligation of saying Mass at the time to which he has agreed. Conversely, a priest whose Mass has not been announced on the previous Sunday from the pulpit

conventualis vel paroecialis in nocte Nativitatis Christi celebranda, non Missa quae ex indulto tunc celebrari potest (can. 821, § 2)."

[12] Wernz-Vidal, *Normae Generales,* n. 248, b, note 19: "Non sensu *liturgico,* sed potius sensu vulgari, sc. quae ex devotione celebratur."

[13] "Ecclesiastical laws must be interpreted according to the meaning proper to the text and context. If their meaning remains doubtful or obscure, recourse should be had to parallel passages of the Code, if there are any, to the end and circumstances of the law, and to the mind of the legislator."—Cicognani, *Canon Law,* p. 608.

or by a notice at the church door has no obligation to the public and may not only determine the hour when he is to celebrate, but also the mode according to which that time is to be computed. Thus, priests on vacation are generally free to determine when they shall say Mass. It would be allowable for them to start Mass just before one o'cock P. M. according to any one of the prescribed computations.

This conclusion is based on the natural and canonical meaning of the word *private*.[14] It is rather difficult to think of a high or solemn Mass as private in this sense, but there is no contradiction in the terms. One could finally add that, according to the wording of the Code, the term *private* applies not to the Mass itself but to the manner of celebrating it. Canon 33, § 1, does not use the expression *privata Missa,* but *privata Missae celebratione.* Thus the *Missa pro populo* is in some respects a public Mass, but it may be celebrated in a private way, in a private oratory for instance.

§ 2—The Private Recitation of the Breviary

The word private in this case is at least somewhat explained by an official decree of the Holy See. The sacred Congregation of Rites considered as private that recitation of the divine office which was not choral—*in publica seu chorali recitatione.*[15]

A few authors adhere to this text and consider the choral recitation of the office as public.[16]

On the other hand many well known authors favor a more liberal interpretation and consider as public only that choral recitation which is imposed by law. Such is the recitation of the

[14] Cf. canon 483, n. 1, where it is said that the local ordinary can, in certain instances, force the rector of a church to hold ecclesiastical functions at stated hours. A Sunday Mass celebrated according to the ordinary's prescriptions would be neither parochial, conventual, nor capitular. Yet one would hardly consider it a private Mass.

[15] S.R.C., *Placentia in Hispania,* 12 maii 1905—*Fontes,* n. 6338.

[16] Thus Toso, *Commentaria Minora,* I, 103; Blat, *Commentarium,* I, n. 95; Ojetti, *Normae Generales,* p. 199; Lacau, *De Tempore,* p. 39, note 1; Cicognani, *Canon Law,* p. 678.

office by a cathedral or collegiate chapter. Thus Vermeersch maintains that, if secular priests not held to the choral recitation decided to say the office in common during a retreat, they would do so freely and could use any one of the accepted time-reckonings mentioned in canon 33, § 1.[17]

Cardenas adds another consideration when he writes: "It is probable that religious who are bound by the obligation of choir (canon 610) may enjoy the concession of canon 33, § 1, if the Office is not sung but recited, or if the chanting takes place outside choir." [18]

If one applied the principles set down in the preceding paragraph, the logical conclusion to draw would be that the choral recitation imposed by law is of its nature public, but that, if for some legitimate reason the divine office is not recited as it should normally be, one could infer that it ceases to be public and that it would enjoy the privilege given to any private recitation. Conversely, the free choral recitation by diocesan priests is of itself a private affair, but it could easily become *de facto* public. The chanting of Sunday Vespers in parochial churches, for instance, would then be considered a public recitation of the divine office. This question is of course more theoretical than practical, since one can hardly imagine such a ceremony close to midnight.

§ 3—The Reception of Holy Communion

This question is of minor importance since it is solved at least implicitly by the proper disposition of another consideration, namely, that of the distribution of Holy Communion. Canon 867, § 4, tells us that Holy Communion may be distributed during those hours when the celebration of Mass is permitted, unless there is good reason to give it at other times. Mass, under or-

[17] Cf. *Epitome,* I, n. 148. This interpretation is admitted by Michiels (*Normae Generales,* II, 137-138), Van Hove (*De Temporis Supputatione,* n. 291), Vromant (*Normae Generales,* p. 164, note 3), Beste (*Introductio,* p. 100), Cardenas ("Midnight and Canon Law,"—*ER,* CI [1939], 403).

[18] *Op. cit.,* p. 403, n. 2.

dinary circumstances, may be begun from one hour before dawn to one hour after noon.[19]

When, for some special reason, Mass is celebrated at some other hour, Holy Communion may then also be received, unless that be specifically prohibited. The permission to receive Holy Communion is explicitly granted for the Midnight Mass on Christmas day.[20]

The practical doubt to solve at this point is the question of one's natural fast before the reception of Holy Communion. Is one able to reckon the midnight for the observance of this precept according to any one of the reckonings of canon 33, § 1? Nearly all the authors hold that the person who intends to receive Holy Communion the following morning may begin his midnight fast according to any one of the recognized computations.

Coronata writes that the Code mentions only the reception of Holy Communion, and that the wording of the law would seem to exclude the eucharistic fast, but that, since the eucharistic fast is closely associated with the reception of Holy Communion, the observance of this precept could seemingly be made according to any one of the reckonings mentioned in canon 33, § 1.[21]

Many authors teach that the list of exceptions in canon 33, § 1, is merely illustrative and not exhaustive. For them the problem now under consideration deserves no special mention.[22] Many of the writers who hold that the enumeration is exhaustive simply look upon the eucharistic fast as included at least implicitly with the reception of Holy Communion or at least with the fourth exception concerning the law of fast and abstinence.[23]

[19] Canon 821, § 1.

[20] Canon 821, § 3. The same may be logically maintained for other midnight Masses celebrated, for instance, during Eucharistic Congresses. Cf. Cicognani, *Canon Law,* pp. 677-678. Special norms for Holy Week are found in canon 867, §§ 2-3. Cf. Jorio, *La Comunione agl' Infermi* (Romae: Pustet, 1931), n. 47.

[21] *Institutiones,* I, n. 51, note 5.

[22] Cf. *infra,* pp. 139-142.

[23] Cf. Toso, *Commentaria Minora,* I, 104; Ojetti, *Normae Generales,* p.

Eichmann is possibly the only author to imply that this prescription of canon 33, § 1, applies only to the hour at which Holy Communion may be distributed, and not to the reckoning of time with reference to the eucharistic fast.[24]

§ 4—The Observance of the Law of Fast and Abstinence

This exception certainly refers to canon 1246, where fast and abstinence are mentioned together. If the eucharistic fast is also included, this is done merely in an implied or indirect way. There seems to be no practical difficulty to solve here.

Article 2—Exhaustive of Illustrative Enumeration

This is another knotty and controverted point, though it is more theoretical than practical. The point at issue is whether or not the four explicit exceptions mentioned in canon 33, § 1, are the only cases for which any other than usual time may be adopted, or whether that privilege is to be extended to the observance of all laws of a private nature, such as the abstention from servile works on Sundays and holy days of obligation and the reckoning of the time granted for the visitation of churches to gain indulgences.[25]

It is highly probable that the number of authors holding that

199; Michiels, *Normae Generales,* II, 138; Vromant, *Normae Generales,* p. 164, note 4; Van Hove, *De Temporis Supputatione,* n. 292.

[24] *Lehrbuch des Kirchenrechts auf Grund des Codex Iuris Canonici* (2. ed., Paderborn: Druck und Verlag von Ferdinand Schöningh, 1926), p. 48.—Cf. Van Hove, *De Temporis Supputatione,* n. 292, note 4.—Michiels (*Normae Generales,* II, 138, note 2) notes that if one were not allowed one's choice for the reckoning of midnight for the eucharistic fast on account of its close association with the distribution of Holy Communion, one would be granted this privilege in virtue of the fourth exception explicitly mentioned, namely, that of fast and abstinence. This author maintains that the word *fast* applies either to the eucharistic or to the ecclesiastical fast.

[25] Cf. canons 1248 and 923. The question of the eucharistic fast could also be added here, but this topic is included with the reception of Holy Communion even by the authors who hold that the enumeration is exhaustive. Cf. *supra,* pp. 136-137.

this enumeration is exhaustive is slightly superior to that of the writers who maintain a different view.[26]

If one adheres to the strict wording of the text of canon 33, § 1, the enumeration is obviously exhaustive. It is indeed explicitly stated that usual time is to be followed for the reckoning of the hours of the day, but that one is free to use either local time, true or mean, or legal time, regional or extraordinary, in the four cases already referred to. The strict interpretation is made stronger by canon 31 which rules that the norms given in canons 32-35 are to be followed in the reckoning of time unless the contrary is expressly stipulated.

Such is the gist of Van Hove's argumentation.[27] He also adds that the liberty of choosing any other computation than the usual one is an exception to the law, which law calls for the use of

[26] A—The strict interpretation is advanced by the following authors: Beste, *Introductio,* p. 100; Chelodi-Bertagnolli, *Ius de Personis,* p. 158, note 3; Cicognani, *Canon Law,* p. 679, Coronata, *Institutiones,* I, n. 51, d; Lacau, *De Tempore,* n. 39, B, b; Maroto, *Institutiones,* I, n. 258, 2; Michiels, *Normae Generales,* II, 138-139; Ojetti, *Normae Generales,* p. 194, note 4; Oesterle, *Praelectiones Iuris Canonici* (Romae: in Collegio S. Anselmi, 1931), 19—(this work will hereafter be cited as *Praelectiones*); Toso, *Commentaria Minora,* I, 103-104; Van Hove, *De Temporis Supputatione,* n. 294; Vromant, *Normae Generales,* n. 133, 2.

B—The more liberal opinion which holds the enumeration to be illustrative in character is represented by: Cappello, *Summa,* I, n. 179, 2, 4; Cardenas, "Midnight and Canon Law,"—*ER,* CI (1939), 410; Claeys-Boúúaert, *Selecta Capita,* p. 9; Claeys- Boúúaert-Simenon, *Manuale Iuris Canonici,* I, n. 191; De Meester, *Iuris Canonici et Iuris Canonico-Civilis Compendium* (3 vols., Brugis: Sumptibus et Typis Societatis Sancti Augustini, 1921-1928), I, n. 286—(this work will hereafter be referred to as *Compendium*); Leroux, "L'Heure à Suivre dans l'Observation des Lois Ecclésiastiques,"—*Revue Ecclésiastique de Liége,* XI (1919-1920), 163; Vermeersch-Creusen, *Epitome,* I, n. 148; Vermeersch, *Theologiae Moralis Principia, Responsa, Consilia* (2. ed., 4 vols., Romae: Universitas Gregoriana, 1926), I, n. 383—Hereafter this work will be cited as *Theologia Moralis.*—Indulgences are said to fall under canon 33, § 1, according to Beringer-Steinen, *Die Ablässe, ihr Wesen und Gebrauch,* as cited by Van Hove (*De Temporis Supputatione,* n. 294, note 3).

[27] *De Temporis Supputatione,* n. 294.

usual time, and is therefore to receive a strict interpretation according to canon 19. He finally mentions that this is confirmed by the fact that the great majority of authors are of this opinion. And one may also add that where the law does not distinguish neither may private individuals distinguish.[28]

The writer nevertheless feels that these arguments are not conclusive. In the first place, if one adheres very strictly to the wording of the canons of Title III, strange contradictions will indeed follow. It has already been shown that canon 32, § 1, cannot easily be understood in the light of canon 34, § 2. Canon 32, §1, states that the day consists of 24 hours, and that these are to be reckoned continuously from midnight to midnight. Conversely, canon 34, § 2, provides for the natural reckoning of time according to which the 24 hours are not necessarily computed from midnight to midnight. One can add that this exception to canon 32, § 1, is by no means expressly stated. It will also be shown in the next chapter that the natural reckoning of time for suspensions is an implicit derogation to the norms given in the very same canon for the civil and the natural reckoning of time.

In the second place it is not too certain that exceptions to law as envisaged by canon 19 are in reality such as are found in the Code itself. Vermeersch maintains that the only exceptions to the laws contained in the Code would be particular laws, or general laws enacted subsequently to the Code and promulgated as formal exceptions to a truly universal law. The Code, indeed, has been promulgated *ad modum unius* and one cannot strictly think of any exception to law within this unit.[29]

In fine it may be stated that the more common teaching of the canonists on any particular point is not in any way a conclusive argument. It means practically nothing when equally noted authors, even though their number be inferior, hold a divergent view.

A very strong argument for the illustrative theory is found in

[28] Cf. Oesterle, *Praelectiones,* I, 19.
[29] *Epitome,* I, n. 126, 3.

the fact that in the pre-Code legislation all private ecclesiastical obligations without exception followed the same rule relative to the reckoning of time. For any kind of private obligation a person could then select any one of the then acceptable computations. The names of these various reckonings were not exactly the same as those mentioned in canon 33, § 1, but they substantially corresponded to them.[30]

One cannot contend with Van Hove that the rescript of August 9, 1899, to Holland no longer has any application now in as far as the use of various reckonings [then not conceded] is now granted.[31] That is not beyond doubt. It might be true to say that the pre-Code authors did not generally consider this possible, but nothing is proved beyond that point. Incidentally, Vermeersch saw before the promulgation of the Code the possibility of using various computations on the same day.[32]

Michiels insists that even if the rescript of 1899 had granted the privilege of using more than one computation on the same day, it would have been revoked by canon 33, § 1, in virtue of canon 6, n. 6. This, however, would seem a rather uncertain means of revoking privileges that certainly existed before the Code.

Furthermore a confirmation of the illustrative theory was given by the Sacred Penitentiary when it approved the classical work on indulgences by Beringer-Steinen, *"Die Ablässe, ihr Wesen und Gebrauch,"* wherein it is held that the reckoning of time in the question of indulgences may be made according to any of the divergent computations which are now conceded for use by the ruling of canon 33, § 1.[33]

It may be said with Van Hove that this is not an apodictical proof since it is not of the competency of this Sacred Tribunal to interpret the general laws of the Church.[34] It does, however,

[30] Cf. *supra*, pp. 97-100.

[31] Cf. *De Temporis Supputatione,* n 294, note 3.

[32] Cf. Creusen, "Minuit Canonique,"—*NRT,* L (1923), 468.

[33] Cf. Beringer-Steinen, *Die Ablässe, ihr Wesen und Gebrauch* (2 ed., 2 vols. and Supplement, Paderborn: Schöningh, 1921-1930).

[34] *De Temporis Supputatione,* n. 294, note 3.

allow one to conclude that the theory which looks to canon 33, § 1, as offering an exhaustive enumeration is not above all doubt, and that in virtue of canon 15 one is free to use the other opinion with a clear conscience.

Other arguments in favor of the illustrative theory may be found. Indeed one cannot think of any valid reason why other obligations of a private nature should be excluded. It would seem logical to apply the same rule to all obligations of the same nature. Canon 33, § 2, seems to be a clear indication that the faculty of choosing one reckoning or another is not restricted to the four institutes explicitly mentioned in canon 33, § 1. If this provision concerning contracts had not been intended as applying merely to the reckoning of the hours, it seems that it should have been placed immediately after canon 31.[35]

The question especially of the abstention from servile works on Sundays and holy days of obligation is not of a great practical importance, since no one is likely to work at midnight unless he has to, and in that case the precept does not bind.

Article 3—The Use of the Various Reckonings

§ 1—The Various Reckonings Permitted by Canon 33, § 1.

Canon 33, § 1, mentions, besides the usual time of the place, local time, true or mean, and legal time, regional or extraordinary.

[35] Cf. Vermeersch-Creusen, *Epitome,* I, n. 148: "Menti legislatoris consentaneum putavimus ut amplam istam facultatem ad indulgentias quoque et requiem festivam traduceremus."—Cf. *supra,* pp. 118-120, concerning contracts. De Meester (*Compendium,* I, n. 286) invokes canon 20 in favor of the illustrative theory. Michiels (*Normae Generales,* II, 139, note 3) justly remarks that the question to solve here is the proper interpretation of the law and not the supplying of any deficiency, which later consideration is the one with which canon 20 is concerned. This is true, in general, but if one admits that the reckoning of time in the matter of gaining indulgences is exceptional, one may logically argue that canon 20 must then necessarily be invoked, since there would not otherwise be any canonical prescription relative to the reckoning of the hours of the day in canon 923, for instance. Without canon 20 no provision of canon 33, §1, would apply, not even the ruling that usual time must be followed.

The two forms of local time have already been explained. True solar time depends on the actual motion of the earth around the sun; a true solar day is the interval of time that lapses between two successive transits of the sun over the same meridian. This time varies throughout the year.[36] On the other hand mean solar time is determined by the passage of a fictitious sun over the same meridian. A mean solar day never varies in length and is nothing else than a fixed average day whose uniform length results from an equalization of the extremes inherent in the true solar days of any given year.[37]

These explanations agree with those given by practically all the canonists who define these terms.[38]

Usual time has a rather loose meaning. It simply refers to the time actually followed in any given place. It generally corresponds to one of the various classifications of time mentioned in canon 33, § 1, namely, to local time, true or mean, or to legal time, regional or extraordinary. An instance of usual time that does not coincide with any of these four possible reckonings is the old Roman system of computing the hours of the day from sunset to sunset.[39] Ships generally have a time of their own which very seldom corresponds to any of the various classifications intimated by canon 33, § 1.

According to some authors there are two kinds of usual time; the scientific and the popular. The former is followed in legal and public transactions, while the latter is used by private indi-

36 Cf. *supra,* pp. 43-46. Cf. also *infra,* pp. 258-263, where tables are given for the computation of true solar time throughout the year.

37 Cf. *supra,* p. 43.

38 Cf. Maroto, *Institutiones,* I, n. 258, 1, A, b; Cicognani, *Canon Law,* pp. 666-669; Michiels, *Normae Generales,* II, 127-129; Van Hove, *De Temporis Supputatione,* nn. 286-287; Jorio, *La Comunione agl'Infermi,* Appendice III, pp. 90 ss.—Cappello, in one of his works (*Summa,* I, n. 177, 2, 2*) interchanges the true definitions in their relation to these two terms of *true* and *mean* solar time. This is undoubtedly due to a distraction, since these same terms are elsewhere properly defined by the author.

39 Cf. *supra,* p. 38; Maroto, *Institutiones,* I, n. 258, 1, A, b; Ojetti, *Normae Generales,* p. 197, note 1; Lacau, *De Tempore,* n. 38, A, d.

viduals in every-day life. The reckoning of the hours of the day may be made according to either of them.[40]

An example of this twofold usual time existed a few years ago in some of the New England States when the use of Daylight Saving Time (DST) was not uniform throughout the States but varied with the municipalities. In many localities the official time was still Eastern Standard Time (EST), but many manufactories ran according to DST, and the great majority of the workers followed this time.

Now, there would be no difficulty in such cases if some of the public clocks indicated DST. That time would then certainly be usual time and could be licitly followed. One might wonder, however, concerning the lawfulness of adopting a time that is more or less usual but nowhere indicated by the public clocks.

Michiels states that usual time is the one *de facto* followed by the greater part of the community and generally indicated by the public clocks.[41] Vidal is of the opinion that the *tempus usuale* is based rather on the usage of the people than upon a scientific computation of time and that it may derive from various causes, e.g., from the difficulty of communication, from the lack of a town-clock, which lack is then supplied for by the ringing of bells at sunrise, at noon, and sunset, and by following the sundial or a private watch.[42] Vermeersch is more specific. He maintains that usual time depends on the custom of the place, and not precisely on the time indicated by clocks.[43]

[40] Cf. Wernz-Vidal, *Normae Generales*, n. 248, a: "Duplex autem simul dari potest tempus *usuale*, v. gr. unum pro negotiis quae publica auctoritate reguntur, aliud pro rebus ad vitam privatam spectantibus *usus communis vulgaris*, et iuxta utrumque horae supputari possunt. Plerumque unicum erit tempus usuale, cum populus soleat sese accommodare horae legali, cuius cognitio plerumque in praxi erít facilior."

[41] *Normae Generales*, II, 139. Cf. Vromant, *Normae Generales*, n. 132, 1; Beste, *Introductio*, p. 100.

[42] Cf. Wernz-Vidal, *Normae Generales*, n. 245.

[43] "Usus loci is est qui *re vera* observatur communiter in loco. Quare minus ad indicationes horologiorum quam ad mores populi attendendum est; et alius esse potest usus in aliis locis eiusdem dioecesis."—*Theologia Moralis*, I, n. 383.

It would therefore seem logical to conclude that DST could be called usual in a community whose legal time is indeed Standard Time, but where the mills run and general business is transacted, during the summer months, one hour earlier than during the rest of the year, so as to correspond to neighboring communities whose official time is DST.

The word *locus* has a very elastic signification, and may apply to a city, a section thereof such as a suburb, or even to a single house or family in some isolated spot.[44]

Legal time is another generic expression and refers to the time that any duly constituted government prescribes at least for one obligation.[45] This government might be either Federal, State or municipal.

There are three kinds of legal time. Two of these are mentioned in canon 33, § 1, i. e., regional and extraordinary. The third is zone time and is considered in an answer given on November 10, 1925, by the Pontifical Commision which was established for the authentic interpretation of the canons of the Code.[46]

The word *regional* as applied to time is new in Canon Law. Some authors are of the opinion that in the mind of the framers of the Code *regional* really meant *zonal,* and that this word was adopted for the sake of precision, since, in the language of astronomers, *zonal* refers to the zones or portions of the earth's circumference enclosed within two parallels, and not to those contained within two meridians, which is the accepted meaning in Canon Law.[47]

[44] "*Locus* est vocabulum indeterminatum quod maiore vel minore extensione sumi potest: locus dicitur pars civitatis, pagum, et etiam universa regio; potest esse conglobatio paucarum domorum,. vel etiam unius familiae habitatio in deserto. Usualis supputatio est illa quae communiter adhibetur in loco, atque potest esse communis universae regionis vel omnino particularis loci angustioris."—Vromant, *Normae Generales,* n. 132, 1.

[45] Cf. *supra,* p. 99; Michiels, *Normae Generales,* II, 139; Van Hove, *De Temporis Supputatione,* n. 286.

[46] Cf. *AAS,* XVII (1925), 582; Bouscaren. *CLD,* I, 59.

[47] Cf. *Periodica,* XIV (1925), 179, 2; *Monitore Ecclesiastico,* XXXVIII (1926), 13.

If that be the case, then indeed one would also have to admit that confusion was rampant at the time of the codification, since zonal time as such is not necessarily legal time, but would have been implicitly considered as such if the subdivision of legal had been zonal and extraordinary, instead of regional and extraordinary, as it now exists in canon 33, § 1.

Regional time as such is distinct from zone time and is the legal time prescribed for a certain region, which might be a whole country or a part thereof. In that sense zone time may be said to be regional time provided it be legal. In some countries there is a clear-cut distinction between the two. Thus, in the Netherlands regional time for the whole country is that of the Amsterdam meridian which is 19 minutes and 32.1 seconds ahead of Greenwich mean time. Yet the Netherlands are within the eastern half of the Greenwich zone, but the time as marked by this zone is not now legally recognized in the Netherlands.

The division of the world into 24 time zones was considered especially by two congresses, held in Washington, D. C., 1884, and in Rome, 1892. The representatives of the various nations were not empowered to legalize zone time for any country. The legality or legal adoption of that time depended on the action of the respective governments, many of which adopted zone time toward the end of the last century. In the United States legal action for the whole country was not taken before 1918.[48] Furthermore, the zones as adopted by the various governments did not coincide exactly with the ones projected by the congresses. Thus, in the United States, Ohio which lies roughly between 80° 45′ and 85° long. West should normally use EST in its eastern half and CST in the remainder of the State. The whole region, with the exception of a few border towns, is actually under EST.[49]

This explanation coincides perfectly with the decision of the Code Commission of Interpretation concerning the use of zone time. It was asked whether or not the so called zone time could be everywhere employed for the purposes mentioned in canon 33.

[48] Cf. *Standard Time Throughout the World*, p. 3.

[49] Cf. map, p. 147, for other instances.

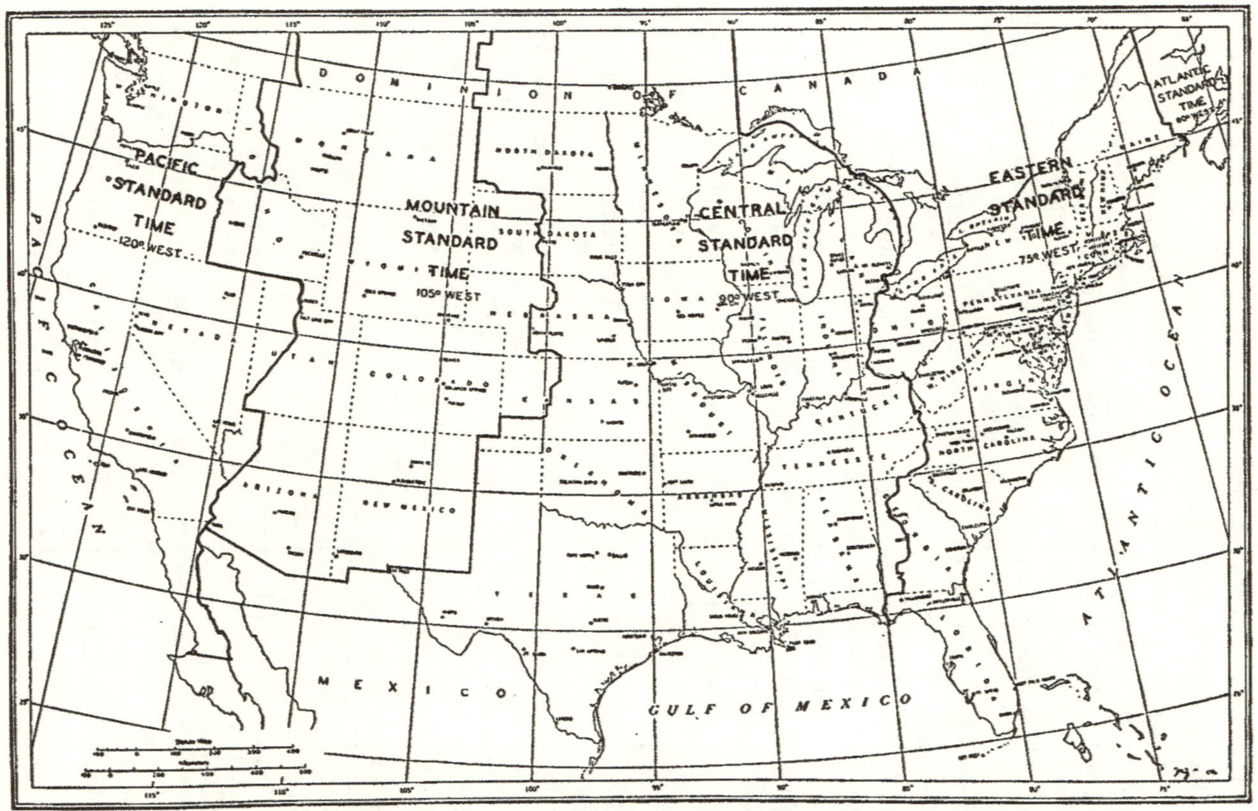

FIGURE 1.—*Standard time zones of the United States, with adjacent parts of Canada and Mexico*

§ 1. The answer was in the affirmative, provided that the zone time is legally recognized.[50]

It is obvious that this decision does not run counter to canon 33, § 1, so as to prevent the use of zone time when it is accepted merely as the usual and not as the legal time. Usual time may be followed everywhere, under any form. The Commission simply wanted to make clear that zone time is of itself not legal until it has been officially adopted by the respective governments and that it cannot be employed as legal time before action to that effect has been taken.[51]

Extraordinary legal time is a special time imposed by law on account of some unusual or accidental cause.[52] In the writer's opinion the best example of such a reckoning occurred during the first World War when the Germans imposed Central European Time on Belgium.[53]

Some authors mention DST as extraordinary legal time.[54] This seems to be only partly true. In many instances DST is determined by statute and goes into effect mechanically from one certain Sunday to another within the year.[55] This change is foreseen

[50] PCI, 10 nov. 1925—*AAS,* XVII (1925), 582: "An ubique terrarum, in casibus canone 33, § 1 expressis, tempus vulgo *zonarium* sequi quis possit? —Affirmative, dummodo hoc tempus sit legale."

[51] Cf. Michiels, *Normae Generales,* II, 140; Wernz-Vidal, *Normae Generales,* n. 248, b. Cf. also Browne, "Zonal Time,"—*IER,* XLIII (1934), 645; *Ibidem,* pp. 420-422.

[52] Cf. Michiels, *op. cit.,* p. 131, 4.

[53] Cf. Cappello, *Summa,* I, n. 179, 2, 2*; Claeys-Boúúaert, *Selecta Capita,* p. 9; Creusen, "Minuit Canonique," *NRT,* L (1923), 465; Coronata, *Institutiones,* I, n. 50, I; Toso, *Commentaria Minora,* I, 104.

[54] Cf. Wernz-Vidal, *Normae Generales,* n. 245; Cicognani, *Canon Law,* p. 680; Vromant, *Normae Generales,* n. 132, 4, b; Van Hove, *De Temporis Supputatione,* nn. 286-287.

[55] Such is the case in Massachusetts and New Hampshire. The use of DST in this country generally ranges from the last Sunday of April to the last Sunday in September. Cf. *The World Almanac* (1940), p. 166.—DST is usually one hour faster than Standard Time, but in some countries the clocks are set ahead only 30 minutes, or even only 20 minutes in some rare instances. Thus, in New Zealand the clocks are advanced 30

and recurs from year to year just as regularly as the month of April itself. If this form of DST were called extraordinary legal time, this expression would become well high meaningless. In the cases just considered DST is rather a second form of regional time.

On the other hand DST is certainly extraordinary legal time when it is prescribed for a special reason that does not occur regularly every year. An instance of this would be found in England where the use of DST was prolonged in the fall of 1939 on account of the war.

The question now arises whether or not Standard Time is to be considered legal when DST is imposed by law. In other words, is one free to follow Standard Time when DST is in force?

In most European countries where there is but one source of authority, the central, there seems to be no difficulty. When DST is made legal and imposed, Standard Time can no longer be used as legal time. Such is the case in France and Belgium. Standard Time may then be followed only if it happens to coincide with true sun time, with mean sun time, or with the time that is in general use among the community as the usual time.[56]

In this country the situation is altogether different, since the State and the municipal governments legislate freely on the reckoning of time. Under the present day arrangement a federal law makes Standard Time legal throughout the year for the

minutes, and on the Gold Coast only 20. Cf. *Standard Time Throughout the World*, pp. 23-24.

[56] Cf. Cicognani, *Canon Law*, p. 682; Cance, *Le Code de Droit Canonique*, I, n. 70, 2, b; Van Hove, *De Temporis Supputatione*, n. 286, note 3, n. 287, note 1; Toso, *Jus Pontificium*, VI (1926), 5-6.—Maroto (*Institutiones*, I, n. 258, 1, C.) is possibly the only author who maintained explicitly that the fundamental regional time of a country could be used throughout the year even when DST was imposed by law. This author, however, wrote before the decision of the Pontifical Commission of Interpretation under date of Nov. 10, 1925. There is no longer any doubt on the matter at present.

whole country. The United States proper are divided into four zones and the Standard Time of each respective zone must necessarily be followed in certain instances:

> Within the respective zones created under the authority of this subdivision of this chapter the standard time of the zone shall govern the movement of all common carriers engaged in commerce between the several States or between a State and the Territory of Alaska and any of the insular possessions of the United States or any foreign country. In all statutes, orders, rules, and regulations relating to the time of performance of any act by any officer or department of the United States, whether in the legislative, executive, or judicial branches of the Government, or relating to the time within which any rights shall accrue or determine, or within which any act shall or shall not be performed by any person subject to the jurisdiction of the United States, it shall be understood and intended that the time shall be the United States standard time of the zone within which the act is to be performed. Mar. 19, 1918, c. 24, #2, 40 Stat. 451.[57]

In the original law there was a provision that the time of each zone should be advanced one hour on the last Sunday in March of each year and returned to normal time on the last Sunday of October, so that during such "daylight-saving period" the Standard Time in each zone should be one hour in advance of that for the degree of longitude governing the zone.[58] This provision was however abrogated the following year. On the last Sunday in October, 1919, national Daylight Saving Time ended.[59]

Within the next few years, however, DST was made obligatory in some States for State and local affairs. Such a law exists today in Massachusetts; it was passed in 1920 and amended in 1921.[60]

[57] 15 *U. S. C. A.* #262. Cf. *supra*, p. 145, where the nature of legal time is determined.

[58] Cf. *Standard Time Bulletin of the ICC*, p. 1.

[59] *Standard Time Bulletin*, p. 2.

[60] Cf. 272 U. S. 525, 71 L. ed. 387, 47 Sup. Ct. Rep. 189 (1926).

In 1925 "opponents of the daylight-saving law of Massachusetts brought suit to enjoin State officials from enforcing the provisions of the State law on the ground that they were in conflict with the Standard Time Act, and therefore unconstitutional. The court held that the Federal act, when confined to the matters specifically covered by its terms, is not exclusive of State action governing local time, and that there is no conflict between the two statutes. Upon appeal, the Supreme Court of the United States affirmed the decree." [61]

The only conclusion from the above is that Standard Time may be followed throughout the year everywhere in the United States. DST may also be employed when it is legal or usual.

In Michigan, for instance, a State law prescribes EST the year round throughout the State.[62] However, only the lower peninsula of Michigan is in the EST zone of the Federal plan. The upper peninsula is still within the CST zone.[63] Consequently there are two legal times in Marquette, for instance, and there one is free to use either EST or CST the year round. A similar situation exists in El Paso, Tex. The whole of Texas falls within the Federal Central Time Zone, but El Paso is actually using Mountain Time. In this community, therefore, Central and Mountain times may be legitimately adopted throughout the year.

In 1919 the Federal line of demarcation between the Central and the Eastern Time Zones ran practically through the center of Ohio, but it has been gradually pushed westward so that the entire area of Ohio is now embraced in the Eastern Zone.[64]

Consequently, a few years ago, the faithful of western Ohio could consider CST as legal and could lawfully adopt it whether it was the usual time or not. This practice is no longer justified unless CST be the usual time.

[61] *Standard Time Bulletin,* p. 4. Cf. Massachusetts State Grange v. Benton, 10 F. (2d) 515 (1925); 272 U. S. 525, 71 L. ed. 287, 47 Sup. Ct. Rep. 189 (1926).

[62] Cf. *Standard Time Bulletin,* p. 2.

[63] Cf. map, p. 147.

[64] Cf. *Standard Time Bulletin,* p. 2.

Before giving a summary of the various reckonings the faithful are allowed to use, the writer would like to say a word about sidereal time. At one time Augustine held that one could adopt sidereal time: "In fulfilling the duties mentioned in the canon [33, § 1], one may follow sidereal time, if one is a good astronomer." [65]

This assertion was undoubtedly due to a distraction on the part of Augustine. Sidereal time, it is true, is a scientific or astronomical entity, but it has no place in Canon Law. No other canonist, to the present writer's knowledge, contends that sidereal time may be used in canonical computations. Furthermore, Augustine himself subsequently ceased to mention the possibility of using this form of reckoning.[66]

This article may thus be briefly summarized:

Reckoning of Midnight may be made according to

a) Usual Time—simply means the time which is in common use in any given community. It is generally indicated by the public clocks and will most likely be one of the following times.

b) Local Time
   1—True—is that which is indicated by the sun-dial.
   2—Mean—depends on the relation of the sun to the local meridian, but is so modified as to be always the same and invariable from day to day.

c) Legal Time
   1—Regional—is the local mean time of a given central meridian as applied to a certain territory, nation or part thereof, in a permanent way.
   2—Extraordinary—is that which is prescribed in special circumstances.

d) Zone or Standard Time—is a semi-legal time, so to say. It connotes an international division of the world into 24 time zones. It is now made legal (regional) time in nearly every civilized country, except the Netherlands perhaps, with certain restrictions however.[67]

It has already been seen that the common opinion before the Code held that the first stroke of the clock marked the end of

[65] *Commentary,* I (3. ed., 1920), p. 118.

[66] Cf. *Commentary,* I (6th revised edition, 1931), p. 118.

[67] Cf. Lacau, *De Tempore,* n. 32, where a slightly different division is given.

the preceding hour.[68] The same doctrine is maintained today by practically every canonist.[69] The present writer knows of but one author who follows a different line of thought. Oesterle here brings into the question the famous distinction between favorable and odious affairs. He teaches that in favorable things the first stroke of the clock marks the beginning of the coming hour, but that in odious affairs the same purpose is served by the last stroke.[70]

With reference to the reckoning of the hours of the day a final remark must here be made. It is commonly held today, just as it was before the Code, that the proper computation of the hours is the physical, and not the mathematical reckoning. In other words, one need not worry about the precise and absolute time as indicated by the motion of the heavenly bodies. It suffices to use a good clock as one's guide.[71]

§ 2—The Double Probability

Now that the various time-reckonings have been examined individually, it remains that their use in concrete cases be considered. The gist of the problem centers about the classical example of whether or not one may eat flesh meat Friday night without breaking one's eucharistic fast for the following day. Thus in Washington, D.C., when it is midnight EST it is only 11:52 P.M. local mean time. Is it then permissible to eat meat at midnight EST Friday night on the assumption that Friday is over according to EST, and to receive Holy Communion the following morning on the ground that Saturday, according to the local mean time reckoning, did not begin before 12:08 A.M. EST?

The intrinsic nature of the problem lies in the use of more

[68] Cf. *supra*, p. 101.

[69] Cf. Van Hove, *De Temporis Supputatione*, n. 286.

[70] "Primus ictus horologii nuntiat sequentem horam; alii aliter sentiunt, incipiendo horam post ultimum ictum. Ergo in favorabilibus incipit hora primo ictu, in odiosis ultimo aut viceversa."—*Praelectiones*, I, 17.

[71] Cf. Cicognani, *Canon Law*, p. 684; Oesterle, *Praelectiones*, I, 17, etc.

than one reckoning at the same time. The question as it now exists is new in Canon Law, but it is closely associated with an older one which was concerned with one's line of conduct when one was confronted with two or more divergent clocks supposed to indicate one and the same time. This problem still exists, and must be examined in order to give a thorough understanding of canon 33, § 1.

Before the nineteenth century, when there was no question of using any other time than the usual, one could have been perplexed at midnight by the divergent readings of two or more clocks that were normally correct. Thus if Cornelius wanted to receive Holy Communion the following morning and entered his home just about at midnight and found to his great surprise that one of his clocks read 11:50 P.M. and the other 12:02 A.M., there was no possibility for him of knowing whether it was then midnight or not. According to the common teaching of the moralists Cornelius could then have followed the slow clock to reckon the beginning of his fast. On the other hand, if that day had been a Friday, he could not have argued that it was lawful for him to eat meat up to midnight of the slow clock on the assumption that Friday was over according to the fast clock. He would then have considered a few minutes as belonging both to Friday and to Saturday. This is, in its essential parts, the problem of the double probability.

This problem still exists today. For instance in Boston, Mass., when DST is not in use, the Standard Time midnight gives one the greatest latitude possible for the beginning of the obligation of fasting. It is the last possible point of time demarcating midnight as permitted in canon 33, § 1. It may be supposed that a person is perplexed by the problem of determining the exact point of midnight according to EST if this particular kind of time-reckoning be the only one that is actually involved in his specific case. If one timepiece reads 11:52 P.M. and another 12:03 A.M., the problem is exactly the same as that which confronted Cornelius. The solution is, in itself, also the very same.

In general the use of a double probability is wrong. The matter is thoroughly treated by Lugo as follows:

Sit duplex sententia probabilis: altera dicit hodie ieiunandum esse; altera non esse hodie ieiunandum, sed cras. Ego hodie amplector secundam sententiam et non ieiuno; cras amplector priman, et dico; iam transiit dies, qua debebat ieiunari, liber sum a ieiunio: eaque via, stante lege certa obligante ad ieiunium, absque peccatò omitterem ieiunium, *quod est absurdum.* Nam superior obligans ad ieiunium reddit etiam illicitam mutationem illam sententiae, quae omnino impedit legis observantiam. Nec simile est in illo, qui mutando locum liberatur a ieiunio, si hodie sit Romae, ubi cras sit vigilia S. Bartholomaei, et cras sit in alio loco, ubi hodie facta fuit vigilia. Hoc, inquam, non est simile, quia illa mutatione loci fit liber a praecepto, quatenus nunquam fuit debitor, eo quod nec Romae nec extra invenerit praeceptum. At vero in nostro casu homo comprehenditur praecepto et lex exigit ab illo ieiunium, . . . Atque ideo non potes absque inobedientia non solvere ieiunium, quum sit certa lex ieiunii te obligans, licet non sit certum tempus pro quo obligat. Aliud simile exemplum esse potest in praecepto audiendi sacrum. Si v. gr. die festo, meridie iam imminente, dicas: amplector sententiam dicentem posse fieri sacrum paulo post meridiem; deinde post meridiem, mutata sententia, dicis: amplector contrariam sententiam, quod non liceat facere sacrum post meridiem, atque ideo liber sum a sacro faciendo. Ecce praeceptum erat certum et debitum certum audiendi sacrum. Quare superior obligans ad sacrum, obligat ad retinendam sententiam semel acceptam, ex cuius mutatione impeditur observatio certi praecepti. . . Denique. . . ex eisdem principiis dicendum est non posse sacerdotem amplecti unum horologium ad solvendum ieiunium praecedentis diei et alterum postea ad inchoandum ieiunium naturale, quod requiritur ad celebrandum die sequenti: sed debere sequi unum et idem horologium pro utroque praecepto. Nam licet praecepta sint diversa, et singulis praeceptis seorsim videatur satisfieri. . . non satisfiet tamen toti legi seu utrique praecepto simul; unde videntur sequi eadem inconvenientia. Nam stante illo duplici praecepto, ego debeo ex complexione utriusque praecepti abstinentiam a cibo post comestionem unicam diei praecedentis seu post refectiunculam serotinam usque ad missam vel communionem diei sequentis; et superior habet ius ex vi utriusque praecepti ad hoc totum, ergo habet ius

> ad prohibendam variationem horologii, ex qua impediatur observantia et redditio totius illius debiti, ad quod certum habebat ius: quod enim id debeatur unico vel duplici praecepto, parum refert.[72]

This has been the more common opinion to the present day.[73] However, the use of the double probability really presents a problem of a special nature when applied to the eucharistic fast. Lugo is very cautious when he specifically mentions this topic, much more cautious, indeed, than many modern authors.[74]

Since the promulgation of the Code this problem has been thoroughly examined by Salsmans.[75] This author naturally rejects the use of a double probability in general.[76] He makes an exception, however, for the eucharistic fast. It is obviously wrong for a person who has the intention of receiving Holy Communion the following morning, to eat meat at 11:52 P.M. Friday night according to one clock, but already 12:02 A.M. according to another clock. It is another thing to say, on the

[72] *De Sacramento Eucharistiae,* disp. XV, sect. II, nn. 50-53, as cited by Ojetti, *Normae Generales,* p. 197, note 2.

[73] Cf. Creusen, "Minuit Canonique,"—*NRT,* L (1923), 466.

[74] Cf. Van Hove, *De Temporis Supputatione,* n. 298. Vermeersch (*Theologia Moralis,* I, n. 382) writes: "Re vera respectus utriusque ieiunii, ecclesiastici et eucharistici, diversi sunt et per se separari possunt. Cur coniungendi essent in diei aestimatione? Desideratur argumentum intrinsecum. Summum illud dici potest: propter materialem utriusque causae affinitatem, rationabilis mens legislatoris est ut coniungantur. Quae ratio satis debilis nobis videtur."

[75] Cf. "De l'Usage Simultané de la Double Probabilité,"—*NRT,* XLIX (1922), 148-150.

[76] "La volonté de tirer parti de la double probabilité est donc coupable . . . Il est évident aussi qu'une telle disposition de volonté est, en outre, opposée à toute droiture naturelle. L'exécution ne serait pas moins coupable en règle générale, parce que, comme Lugo le montre très bien sur un grand nombre d'exemples, on violerait certainement une loi ou le droit certain du prochain, et il y aurait là, en plus, une déloyauté indigne d'un moraliste ou même d'un simple chrétien. . . Nous disions en règle générale; car, quand il s'agit du jeûne eucharistique, nous nous demandons si vraiment l'usage de la seconde probabilité reste interdit à celui qui a fait usage de la première."—*Op. cit.,* p. 149.

other hand, that this person has really broken his fast. Thus, if one at the time of eating meat, under the circumstances described above, has not the intention of receiving Holy Communion, one is perfectly justified in so doing. It may now be supposed that, a short time later, this person decides to receive for a special reason, for instance, because he recalls that he had promised to receive with a group, or for any other similar reason.

In this case there is absolutely nothing to prevent one from going to Holy Communion. One has not sinned in using the double probability. This person simply considers the fact that he is not sure of having broken his fast. The situation is the same as that of a person who takes food during the night without thinking of the time, and then wonders the next morning whether it was before or after midnight. Here the common teaching holds that Holy Communion may be received because the midnight fast has not been certainly broken.[77]

On the other hand if one ate meat with the express intention of receiving the following morning under the assumption that it was Friday according to one clock and Saturday according to another, one would certainly be sinning. But even then it cannot be said that the eucharistic fast has been broken. Salsmans concludes that, in such a case, if the person were really sorry and confessed his sin, he would then be free to receive.[78]

Vermeersch, examining the problem from another angle, judiciously admits without restriction the use of the double prob-

[77] Cf. St. Alphonsus, *Theologia Moralis,* lib. VI, n. 282; Lehmkuhl, *Theologia Moralis,* I, n. 205;

[78] "Déjà plusieurs auteurs ne defendent pas la Communion, *si l'intéressé n'a pas voulu simultanément user des deux probabilités,* si Titus par exemple, avait pris de la viande en renonçant à la Communion, et se fut aperçu plus tard que d'après l'autre montre il est probablement encore à jeun. Cfr. Lehmkuhl, 11e ed. Vol. I, n. 205; Genicot, 9e ed. Vol. I, n. 81 b. ad fin. Mais, dans l'hypothèse du texte, le fait que Titus a voulu profiter de la double opinion, peut-être sans faute formelle, change-t-il essentiellement le cas? Noldin admet bien (Vol. I, n. 224) qu'on n'est pas exclu de la communion par le fait qu'on a rendu insoluble un doute au sujet du jeûne eucharistique."—*Op. cit.* p. 149, note 1.

ability wherever distinct obligations are involved. Thus, one could eat meat when it is 12:02 A.M. according to the fast watch on the assumption that it is no longer Friday, and recite Compline for Friday because it is only 11.52 P.M. according to the slower timepiece.[79]

In fine one may conclude in accordance with the traditional teaching of the moralists that it is sinful to make use of the double probability. Such is the case when on Friday night, a person is in doubt as to the precise moment of midnight, and nevertheless eats meat with the intention of receiving Holy Communion the following morning. This, however, does not necessarily break one's fast and the element of sin is the only impediment to the reception of Holy Communion.

On the other hand, when a person makes use of only one probability, no sinful act is posited and the agent is absolutely free to receive Holy Communion. Such is the case when a person, confronted with two divergent clock readings relative to the moment of midnight, decides not to eat meat, but only to take other food up to midnight according to the slower timepiece. Other situations might be thought of. Thus, if Sempronius, who had the intention of receiving the following morning, saw only one clock according to which it was already after midnight, decided not to receive and ate a ham sandwich, but found to his surprise that it was not yet midnight according to another clock, Holy Communion could then be received. In this case no sin has been committed and the fast has not been certainly violated.

§ 3—When Two or More Reckonings are Involved

One is now to leave in the background the case in which two clocks, though actually marking two divergent points of time, are nevertheless supposed to indicate one and the same kind of time, and to consider the simultaneous use of various time-reckonings. The point at issue is to know, for instance, whether or not it is possible to eat meat without breaking one's eucharistic fast. when

[79] *Theologia Moralis,* I, n. 382.

by Daylight Saving Time it is midnight on Friday night, but only 11:00 P. M. by Standard Time. The question of divergent clocks is not here considered. One timepiece suffices.

A. *The Various Opinions*

Many are the opinions given on this point. The strictest one is that proposed at one time in the *Ami du Clergê.* Its contention is that the liberty of using any one of the various reckonings mentioned in canon 33, § 1, is to be interpreted in such a way that when one computation has been adopted it must be followed for the observance of all one's obligations for that day. Thus, if one chooses mean sun time for the reckoning of time relative to the recitation of one's office, one must also use the same computation with reference to the observance of the natural fast, etc.[80]

The reasons for this stand are the following. In the first place any other interpretaion would invariably lead one to consider a certain period of time as belonging simultaneously to Friday and to Saturday, which is absurd.[81] In the second place the problem is said to be the same as that of the double probability, and consequently, one may use any one of the permissible reckonings (*disjunctive*) but not more than one at the same time (*conjunctive*).[82]

This is explained by the so called canonical midnight, which is the coincidence at the middle of the night of two precepts, one of which ceases to exist the very moment when the other begins.[83] This canonical midnight must be strictly observed because the universal usage of the Church admits of no hiatus between two successive days and, above all, because such has been the constant ecclesiastical tradition.[84]

[80] XXXIX (1922), 637; XL (1923), 201.

[81] *Op. cit.* XXXIX (1922), 637.

[82] Cf. *op. cit.,* XL (1923), 200.

[83] "On parle de minuit légal, de minuit solaire, de minuit moyen, etc., Si nous parlions un peu aussi du *minuit canonique?* Nous désignons sous cette appellation nouvelle la *coincidence,* au milieu de la nuit, de deux préceptes dont l'un cesse à l'instant où l'autre commence d'obliger." —*Op. cit.* XL (1923), 200.

[84] "Mais, surtout, ce qui nous semble particulièrement grave dans

Another argument is drawn from the fact that those who hold a different opinion still maintain that food cannot be taken between two consecutive ember days. The juxtaposition of two such days is merely accidental, it is said, and if the opponents were logical, they would allow one to use, for instance, mean time in Boston for the reckoning of the first day, and regional time (EST) for the second day. This would give an interval of sixteen minutes pertaining neither to Friday nor to Saturday. And during this time-interval one could take food without infringing the law.[85]

Other authors, although not admitting the full freedom of choice between the various reckonings so as to allow one to eat meat after midnight of Friday according to one reckoning and yet to receive Holy Communion the following morning because it was not midnight according to another computation, give canon 33, § 1, a wider interpretation than is found in the *Ami du Clergê*. Thus Cappello allows the use of one reckoning for the reception of Communion and another for the recitation of the breviary.[86] Ojetti is not altogether precise and simply states that the absurd must be avoided.[87]

Other canonists are much clearer and state explicitly what is not allowed by canon 33, § 1. They maintain that one is free to

cette question, c'est la constante tradition ecclésiastique et théologique, vivante dans la pratique universelle des fidèles, qui a toujours regardé la fin de l'abstinence comme coincidant exactement avec le début du jeûne eucharistique du lendemain, qui a toujours, par conséquent, admis *'l'unicité'* de ce que nous appelons le minuit liturgique ou canonique, indépendant des variations possibles des horloges."—*Op. cit.*, XL (1923), 201.

[85] Cf. *op. cit.*, XL (1923), 202.

[86] Applicatio huius favorabilis principii tunc solum excluditur cum agitur de actibus seu obligationibus quae *eodem tempore* urgeant; quia nemo potest se eximere citra culpam a certa obligatione quae urget *hic et nunc*, nisi iusta causa excuset." *Summa*, I, n. 179, 2, 3. A similar doctrine is held by Claeys-Boúúaert-Simenon (*Manuale Iuris Canonici*, n. 191, p. 101, 2). They maintain that two or more reckonings may be used unless such an action implies contradiction, or prolongs the time within which certain actions are to be performed.

[87] *Normae Generales*, p. 197, note 2.

choose any one of the lawful reckonings, but not more than one computation for any single act.[88] Maroto, for instance, refers to the classical example, but does not approve the eating of meat on Friday night when it is already midnight according to one reckoning, if one wishes to receive Holy Communion the following morning, but he allows one to use one computation for the breviary and another for the eucharistic fast.[89] A similar view is held by Lacau,[90] McHugh,[91] Toso,[92] Teodori,[93] Woywod.[94]

On the other hand many well known authors draw a more logical conclusion from canon 33, § 1, and maintain that it is altogether permissible for one to use any of the various reckonings in the case of the classical example, that is to say, that one is free to reckon Friday's abstinence according to mean time in Boston for instance, and to compute the beginning of one's eucharistic fast on Saturday according to Standard Time. In other words, a person is allowed to eat meat between 11:44 P. M. and 12:00 midnight EST Friday night and receive Holy

[88] "Licet modo unum sequi, modo aliud; sive altero die unum, altero die aliud; sive etiam in prima materia unum, in secunda, tertia, quarta aliud, etc.; dummodo tamen *unicus actus* non urgeatur ad duas simul materias vel effectus obtinendos, quorum unus et alter diversum tempus exigerent."—Maroto, *Institutiones,* I, n. 258, 2, B.

[89] Cf. *loc. cit.*

[90] *De Tempore,* n. 44.

[91] *HPR,* XIX (1918), 171-172.

[92] *Commentaria Minora,* I, 104.

[93] "De Temporis Supputatione,"—*Apollinaris,* III (1930), 617-619.

[94] *A Practical Commentary on the Code of Canon Law* (4. ed., 2 vols., New York: Wagner, 1932), I, n. 23. This work will hereafter be cited as *Commentary.*

Michiels (*Normae Generales,* II, 141) mentions the following authors: Leitner, *Handbuch des katholischen Kirchenrechts,* I, 16-17; Hilling, *Die allgemeinen Normen,* p. 153, III; Leroux, "L'Heure à Suivre dans l'Observation des Lois Ecclésiastiques,"—*Revue Ecclésiastique de Liége,* XI (1919-1920), 164.—Van Hove (*De Temporis Supputatione,* n. 295, note 1) mentions Niebecker, "Die Berechnung der Tagesstunden (Zeitpunkt der Mitternacht usw.) nach dem can. 33 des C.I.C.,"—*Theologisch-praktische Quartalschrift,* LXXXIII (1930), 774-776. This publication will hereafter be cited in the abbreviated form *LQS.* Cf. also *ER,* LXXXVII (1932), 634-635.

Communion the following morning without infringing either law, namely, that of abstinence for Friday or that of the eucharistic fast for Saturday. This opinion is maintained by Vermeersch,[95] Beste,[96] Blat,[97] Cance,[98] Cardenas,[99] Cocchi,[100] Coronata,[101] Chelodi-Bertagnolli,[102] Creusen,[103] De Meester,[104] Fray Luis de S. Teresa,[105] Genicot-Salsmans,[106] Mitchell,[107] Michiels,[108] Herrera,[109] Van Hove,[110] Vidal,[111] Vromant,[112] and Browne.[113] One could also mention various anonymous writers in the Ecclesiastical Review[114] and the Irish Ecclesiastical Record.[115]

95 *Epitome,* I, n. 148, and especially, *Theologia Moralis,* I, n. 383, quaeritur 2, p. 358 ss.

96 *Introductio,* p. 102.

97 *Commentarium,* I, n. 95.

98 *Le Code de Droit Canonique,* I, n. 71, 3.

99 "Midnight and Canon Law,"—*ER,* CI (1939), 404, 3.

100 *Commentarium,* I, n. 95.

101 *Institutiones,* I, n. 51, c.

102 *Ius de Personis,* p. 159, note I.

103 "Minuit Canonique,"—*NRT,* L (1923), 464-474. Cf. p. 470 where Creusen refers to Mercier.

104 *Compendium,* I, n. 286.

105 "De Re Canonica,"—*El Monte Carmelo,* XXXI (1927), 466-470.

106 *Institutiones Theologiae Moralis,* I, n. 81.

107 "What Time is it? Midnight and Fasting,"—*ER,* LXXXIV (1931), 491-500.

108 *Normae Generales,* II, 142-149. This author (p. 142) refers to Eichmann, *Lehrbuch des Kirchenrechts,* p. 44, note 4; and Ferland, *Semaine Religieuse de Québec,* XXXV (1923), 280-283, 294-298, 567-570, 600-605, 614-621.

109 *Legislacion Eclesiastica sobra el Ayuno y la Abstinencia,* The Catholic University of America, Canon Law Studies, n. 92, (Washington: The Catholic University of America, 1935), p. 123.

110 *De Temporis Supputatione,* nn. 296-299.

111 Wernz-Vidal, *Normae Generales,* n. 248 (p. 386, note 19).

112 *Normae Generales,* n. 133, 2.

113 "Zonal Time,"—*IER,* XLIII (1934), 645.

114 Cf. "Reckoning Midnight for Divers Obligations,"—*ER,* LXXXVI (1932), 532; *ibidem,* LXXXVII (1932), 635-642; "Computing Time according to Canon Law,"—*ER,* LXXXVII (1932), 176.

115 "Zonal Time,"—*IER,* XLIII (1934), 420-422. Cf. also *ibidem* XXVII

The present writer adheres to the opinion set down by these authors not only because it corresponds to the exact meaning of the text of canon 33, § 1, but also because this interpretation is in no way contradictory and absurd, as is evident from the refutation of the arguments of the opponents.

B. *The Solution of the Problem*

The first thing to consider in any law is the text itself, and no one may depart from it if it be clear in itself, unless it be in contradiction, implicitly or explicitly, with another equally clear text. This rule is deduced from canon 18.

An evident case of a law clear in itself but irreconcilable with another clear text is that of canon 32, § 1, where it is ruled that the 24 hours of the day are to be reckoned in continuous succession from midnight to midnight. This conclusion cannot be applied when the natural reckoning is used, since the starting point for the computation of the 24 hours may then be any moment of the day. The only logical interpretation of canon 32, § 1, is that the reckoning of the day from midnight to midnight holds in general, but is not to be followed when other prescriptions of the Code specifically provide for another form of computation.

The case of canon 33, § 1, is similar to that of canon 32, § 1. Certain restrictions are to be found here and there stated explicitly throughout the Code, while others are the logical conclusions or deductions from the consideration of various canons and canonical principles. But in no way is one to limit the prescriptions of this canon on account of preconceived ideas or of a false sentimentality The question is a purely juridical one and must be solved according to juridical principles. It is admitted by every one that the liberal interpretation, even though it seemed irrational in itself, would be perfectly acceptable if the Church so willed it.[116]

(1926), 77; XXXVII (1931), 639.—Cf. *HPR,* XXIII (1923), 1301. Many other authors could be cited.

[116] One reads in the *Ami du Clergé* (XL [1923], 201): "A la vérité, pour qu'il en fût ainsi [that is, to admit the possibility of choosing

The solution of the problem is associated with the determination of the Church's will in this matter. One may say with Creusen that the Church grants the privilege of using more than one reckoning for various obligations on the same day, even when they involve only one act as in the classical example, provided 1—the Code authorize such a line of conduct without restriction and without distinction; 2—provided no contradiction be involved, and 3—provided such an interpretation be not opposed or contrary to the constant and authorized interpretation of a similar law that existed before the Code.[117]

It can now be clearly demonstrated that no such limitation of canon 33, § 1, exists:

1—In the first place it is obvious that canon 33, § 1, itself makes no distinction, and no restriction. There is absolutely no question of using the various reckonings *disjunctive* or in any similar way.[118]

2—In the second place it is beyond doubt that the liberal interpretation is neither contradictory nor irreconcilable with other prescriptions of the Code. There would be contradiction if the simultaneous use of more than one reckoning really implied that a cretain period of time belonged at the same time to Friday and to Saturday, as is said by the opponents of the liberal interpretation.

To solve the case certain suppositions must necessarily be kept in mind when time is considered. Thus, it cannot be said in a most absolute way that Friday ends and Saturday begins when clocks strike midnight. It is admitted by every one that a limitation as to place must be made. When it is midnight in New York it is only 11 P. M. in Chicago. The following hour belongs to Saturday in New York, but is still a part of Friday in

various reckonings for different obligations on the same day], un miracle ne serait pas nécessaire: la volonté de l'Eglise y suffirait. . . Après tout, en pareille matière, les arguments de raison sont sujets à caution: c'est affaire d'autorité, nous ne l'oublions pas."—Cf. Michiels, *Normae Generales*, II, 142, note 2.

[117] "Minuit Canonique."—*NRT*, L (1923), 471.

[118] Cf. Creusen, *op. cit.*, p. 472.

Chicago. It would indeed seem strange to admit a situation in which the very locality allows of various time-reckonings and in which there would nevertheless have to be ruled out the very possibility of pressing such various time-reckonings into service. Thus, if Claudius eats meat Friday night in Boston when it is 11:44 P. M. EST, he does not do so because it is only 11:44 P. M. EST, and therefore still Friday, but because he has chosen mean sun time for the reckoning of the 24 hours during which he must abstain from meat. And at 11:44 P. M. the 24 hours have certainly expired in Boston; Friday no longer exists according to the chosen reckoning. It is obvious that Claudius is under no obligation to abstain from meat for the reason that it is still Friday.

On the other hand there is absolutely no obligation for Claudius to begin his eucharistic fast at that moment since it is not yet Saturday according to EST. When a person eats meat Friday night shortly after midnight (mean sun time, in Boston for instance), and has the intention of receiving Holy Communion the following morning, he does not affirm that it is midnight and not midnight. This would be absurd and contradictory, but in the cited example it is not the case.[119] There would be contradiction, however, if one made use of two divergent clocks to reckon the last possible point of time for midnight so as to eat meat when one clock reads 11:55 P. M. and the other 12:05 A. M. This problem was examined above and is altogether different from the one that is now under consideration.

Another form of contradiction would exist if the simultaneous use of more than one reckoning actually impeded one from completely observing any one of the obligations for which a different reckoning has been chosen.

It is a well known fact that the concurrence of various precepts is of two kinds. There is the concurrence of two precepts that are intrinsically united of their own nature. Such is the concurrence of ecclesiastical fast and abstinence during lent. According to the common teaching of the doctors, these pre-

[119] Cf. Creusen, "Minuit Canonique,"—*NRT,* L (1923), 472.

cepts are so united that they cannot be separated and that the same reckoning must necessarily be used for the two of them. Otherwise one of these precepts would not be observed through 24 hours.

On the other hand there is the concurrence of precepts of a totally different nature, such as the recitation of the breviary and the observance of the eucharistic fast. These are to be considered in themselves individually as independent, separate obligations. There is also another form of concurrence substantially the same as the one just mentioned. It is that which exists when two precepts of a distinct nature are *de facto* joined together in such a way that the same act affects the two precepts. Such is the case when the eucharistic fast for Saturday concurs with the observance of the Friday abstinence. This concurrence is not intrinsic in any way, but merely exterior and accidental. The fast and abstinence of canon 1246 can never *de facto* be separated, while the eucharistic fast and the Friday abstinence are generally separated. Thus the eucharistic fast for any other day in the week than the one which stands in immediate conjunction with Friday has no relation whatsoever with the Friday abstinence.[120]

The simultaneous use of two or more reckonings does not imply contradiction when the connection between the precepts involved is merely extrinsic and accidental. This is proved by the fact that on Monday, for instance, one is free to choose any one of the reckonings, and on Friday, any other. This is admitted by the writer in the *Ami du Clergé,* provided the reckoning chosen for one obligation be adopted also for the others. Therefore, if Sempronius recites his office on Monday according to mean time he can observe his Friday abstinence according to Standard Time, but if one be logical, one cannot say that the reason why Sempronius is justified in so doing is because it is Monday in one case and Friday in the other. The true and only canonical reason is that this priest recites Prime

[120] Cf. Michiels, *Normae Generales,* II, 145; Van Hove, *De Temporis Supputatione,* n. 298.

for Monday at 11:44 P. M. EST Sunday night, which corresponds to midnight Boston mean time, because canon 33, § 1, permits him to do so. A similar conclusion holds for the Friday abstinence. Now, there is absolutely no change in the nature and the fulfilling of these obligations and no change in the law if the choice of mean time for the recitation of the office on Thursday and the choice of Standard Time for the Friday abstinence happen to concur. This is considered legitimate by practically every author, with the notable exception of perhaps only the writer in the *Ami du Clergé*.[121]

The same conclusion must be logically admitted if the two precepts involved embrace one and the same act. If one is able to reckon midnight on Monday according to Standard Time for the eucharistic fast, and the 24 hours of Friday according to mean sun time for the observance of the law of abstinence, there is nothing to prevent one from keeping these two reckonings in the same manner for the same obligations on Friday night entering Saturday. The problem is substantially the same and the most that can be said against the simultaneous use of the two reckonings in that case is that it seems strange. If that be the criterion according to which the whole problem is to be solved, the same argument holds perfectly well also for rejecting the opinion propounded by the writer in the *Ami du Clergé,* since it also leads to strange consequences. One of these is brought out by Vermeersch when he proposes the case of Titus and Caius, two subdeacons, who take food at the same time (midnight marking the end of a vigil). Titus, however, takes flesh meat, while Caius chooses eggs. Now, according to the opinion evolved by the writer in the *Ami du Clergé,* Titus cannot receive Holy Communion the following morning, and he cannot validly finish the recitation of his office. Caius, on the other hand, is not only able to receive Holy Communion the following morning, but he is also qualified to finish his breviary validly. A stranger situation could hardly be thought of.[122]

[121] Cf. Van Hove, *op. cit.,* n. 298.
[122] Cf. *Theologia Moralis,* I, n. 383.

Other strange conclusions follow. In Marquette, Mich., the Federal legal time is CST, but the usual time is EST. Father John, curate in one of the parishes there, generally followed CST for the recitation of his office, and did so up to the vigil of the feast of the Nativity, not thinking that he would be called upon to sing the midnight Mass at which the pastor was to officiate. Now it so happens that the pastor is indisposed at the last moment and calls upon his assistant to sing the Mass. Well, if one follows the opinion of the writer in the *Ami du Clergé,* Father John, even though his office were finished, would have to wait at least for an hour, that is, for midnight CST, before beginning his Christmas Masses. If this is not strange the word has lost much of its original meaning.

3—The third objection to the liberal interpretation would consist in the fact that it is oposed to the constant and universal teaching of theologians and canonists. This is simply denied since it has been shown above that the problem as it now exists in canon 33, § 1, did not arise before 1905. Furthermore, it is merely accidental that the canonists did not consider it as such before the Code, with the exception perhaps of Vermeersch only. And the fact that the question was not considered by canonists does not by any means justify the assertion that they were opposed to it.[123]

It has also been said that the liberal interpretation of canon 33, § 1, would cause scandal among the faithful in whose estimation the end of the Friday abstinence always coincided with the beginning of the eucharistic fast for Saturday. This objection is easily rejected. Indeed, the question of scandal does not in the least affect the law itself, but it might in some instances justly prevent one from *using* what the law permits, according to these words of the Apostle: "All things are lawful to me, but all things are not expedient."[124]

[123] Cf. Creusen, "Minuit Canonique,"—*NRT,* L (1923), 473; Michiels. *Normae Generales,* II, 142-145; Van Hove, *De Temporis Supputatione,* n. 297.—The pre-Code authors considered only the case of the double probability, and that of the lawfulness of using local time in general, even after the public clocks began to indicate regional time.

[124] I Cor., VI, 12.

The conclusion to draw from this point is that if and when the use of the privilege granted in canon 33, § 1, really caused scandal, it would not then be proper for one to make use of the various reckonings. But even this is hard to conceive. For, as Creusen notes, a layman might find this law strange to begin with, but he will gradually understand it if it be properly explained. Furthermore, if one desired to suppress in Canon Law everything that occasions wonderment among the faithful, one would have to delete from the Code a substantial number of canons of Books IV and V.[125]

Another argument of lesser importance is given by opponents to the effect that the so called liberal interpretation is nothing more than casuistry which evades one law, but then necessarily falls afoul of the other. This contention, to have any value, must be not merely stated; it must also be proved. It has been shown conclusively that the eating of meat on Friday night when it is already midnight according to one reckoning and not yet midnight according to a different computation violates neither the law of abstinence nor that of the eucharistic fast. It is equally certain, then, that it does not violate canon 33, § 1, since this would be done only if one violated one or the other of the two laws here in question.[126]

§ 4—Other Problems

A—The Case of the Same Obligation for Two or More Successive Days

In the writer's mind the strongest argument against the liberal interpretation as it has thus far been propounded is the refusal of the authors who hold the liberal interpretation to allow its corresponding application in the case of two successive ember days. Of all the canonists that consider as permissible the eating of meat between Friday and Saturday in the instances already referred to, no one admits the lawfulness of taking food between two successive ember days.[127]

[125] Cf. Creusen, "Minuit Canonique,"—*NRT*, L (1923), 474.

[126] Cf. Creusen, *op. cit.*, p. 372.

[127] Cf. Creusen, *op. cit.*, p. 473; Cardenas, "Midnight and Canon Law,"

It nevertheless seems logical to argue that it is permissible to choose mean sun time for the observance of the fast and abstinence prescribed for the first ember day. (Friday, for instance), and when midnight according to that reckoning has been reached, wait a few minutes so as to compute the beginning of the second ember day (Saturday) according to a different time. In Boston the use of local mean time for Friday and of EST for Saturday would then give one a latitude of 16 minutes during which, as it were, some kind of a common or neutral time is created in the sense that the eating which is not done on Saturday (EST) need also not be considered as being done on Friday (local mean time). The observance of the Saturday fast (by not eating on Saturday, EST) does not imply the taking of food on Friday (local mean time), nor does the observance of the Friday fast (by eating only after midnight, local mean time) imply the taking of food on Saturday (EST). The writer in the *Ami du Clergé*. opposed though he be to any hiatus between two successive ember days, points out that the canonists who admit the liberal interpretation should also accept, if they wish to be logical, the possibility of taking food between two ember days.[128]

The present writer admits the contention found in the *Ami du Clergé,* namely, that if one holds the liberal interpretation, one must logically admit the possibility of taking food between two ember days. For, if in the case of the classical example often referred to, one is able to eat meat Friday night on the ground that the connection between the Friday abstinence and the

—*ER,* CI (1939), 407; Michiels, *Normae Generales,* II, 146; Van Hove, *De Temporis Supputatione,* n. 300; Vermeersch, *Theologia Moralis,* I, n. 383, quaeritur 2.

[128] Cf. XL (1923), 202. The expression *neutral time* is adopted rather reluctantly by the present writer. These same words were at one time used, but later rejected, by Vemeersch: "In priore editione tempus istud vocavimus *quasi neutrale,* utpote non exclusive pertinens ad alterutrum diem, sicut territorium commune inter duas Gentes interdum neutrale appellatur. Sed, quia perperam videtur a nonnullis intellectus esse, isto modo dicendi qui non erat necessarius abstinemus."—*Theologia Moralis,* I, n. 383, quaeritur 2, b. It must however be noted that the expression *neutral time* is in itself perfectly acceptable.

Saturday fast is merely extrinsic and accidental, the same conclusion must also apply to the taking of food between two ember days, since the connection between these two days is merely accidental. The law does not prescribe fast and abstinence for a continuous period of 48 hours which happen to coincide with Friday and Saturday, but for Friday and for Saturday as two separate days, which stand juridically connected only when the same time-reckoning is used in reference to both.

Michiels, who admits the liberal interpretation,[129] but not the possibilitty of taking food between two successive ember days,[130] thus examines various possibilities when obligations are to be fulfilled on consecutive days. He distinguishes between positive and negative precepts.

A positive precept prescribes the positing of a certain act withn a specfied time. Thus, the obligation of reciting the canonical hours binds during the 24 hours that constitute a day, and this day is reckoned from midnight. But it is equally true that the whole day is not necessarily to be used for the fulfilling of this positive obligation. Thus one is free to recite one's breviary within six or seven hours. When that has been done the precept no longer binds for that day, and if one's office is finished some time before midnight nothing prevents the priest from choosing a different reckoning for the following day even though this new computation trespassed on the previous day. Such is also the case when a precept binds only for a part of the day, as the eucharistic fast, for instance.

If Father X, for instance, recites his office for Monday, July 11th according to Standard Time, he is free to begin the office for the following day, Tuesday, July 12th, according to DST, provided Monday's office is finished, even though the 24 hours granted for the fulfilling of the obligation have not yet fully lapsed. On the other hand, if Father X's office for Monday is not finished before midnight he cannot use DST for Tuesday's obligation.[131]

[129] Cf. *supra*, p. 162.

[130] *Normae Generales,* II, 146, 2.

[131] Cf. Michiels, *Normae Generales,* II, 146, 1: "Si ergo hodie re-

However, when the 24 hours have lapsed the obligation ends whether or not it has been fulfilled. If the recitation was partly omitted through neglect, the fault is irreparable in the sense that the obligation can no longer be validly fulfilled. The recitation made after the 24th hour would be merely an act of private devotion. Likewise if a priest through no fault of his is unable to finish the office within 24 hours he is not bound to finish it afterwards, even though the last possible point of time for the reckoning of midnight has not yet been reached. Such a recitation would also be merely an act of private devotion.[132]

The negative precept on the other hand presents a totally different problem. A negative precept forbids the doing of a certain thing for a specified time. Such is the precept of fast and abstinence which necessarily binds through 24 hours (*semper et pro semper*). When two or more days of abstinence follow one another in continuous succession it becomes obvious that the reckoning for the second day cannot trespass on that of the first day for the simple reason that the abstinence of the first day would not then have been observed through the 24 hours that are prescribed by the law. It is not so clear, however, that one is not free to choose for the second day a reckoning that does not coincide with that of the first day, but one that indicates a later time in its computation. For instance, such would be the case if one chose DST for the first day and Standard Time for the second. When the Standard Time day begins the preceding DST day has been over for an hour.

Michiels does not agree with this but thus interprets this problem:

> *Si agatur de praecepto negativo,* seu de actu quodam, qui tamquam "onus diei" per duo vel plures dies *con-*

citationem Matutini incepi media nocte, juxta tempus locale medium computata, non possum omittere Completorium et incipere Matutinum diei sequentis media nocte proxima, juxta tempus legale regionale, (quod supponitur v.g. 15 minutis antecedere horam localem mediam), computata, quippe cum ad finiendam obligationem Completorii adhuc supersint 15 minuta."

[132] Cf. *Ami du Clergé,* XL (1923), 201; *IER;* XXVII (1926), 77-80.

> *tinuos* est omittendus, v.g. de jejunio quadragesimali, luce clarius est, quod *non permittitur* systematis horarii *variatio,* sed supputatio temporis ad incipiendam obligationem libere admissa indubitanter retineri debet ad eamdem finiendam; ratio est, quia in hoc casu obligatio imposita non est onus variorum dierum separatim inspectorum, sed onus, quod ex intentione legislatoris per spatium plurium dierum continuorum, ideoque ad normam can. 35, sine ulla interruptione est ex se sequendum, ita ut omnino excludendum sit etiam minimum temporis praescripti spatium, extra obligationem (jejunium vel abstinentiam) collocatum.[133]

Various cases could be imagined according to which two fast and abstinence days would be continuous *de facto* but not *de iure.* Thus one may suppose the existence, due to a vow, or owing to a sacramental penance, etc., of a particular personal obligation of fasting and abstaining on a certain day. This person would certainly be free to choose any one of the permissible reckonings. Now, if the Wednesday of ember week came along a few days later, it would still be clear that the same person could choose the same or even a different reckoning. To illustrate further, if this special fast and abstinence day happened to fall on the Tuesday of an ember week one could hardly say that the observance of Wednesday's obligation had to begin necessarily when that of Tuesday ended. These two days are certainly continuous *de facto,* but they are nonetheless separate juridical units, and when the person has fasted and abstained through 24 hours for Tuesday and 24 hours for Wednesday the two precepts have been fulfilled and the law demands nothing else.

It cannot now be said that the fact that two continuous days of fast and abstinence which are imposed by the law as continuous *de facto* are also to be considered as juridically continuous, so that no interval of time, however short it might be, can be wedged in between the two. It might be objected that this interval would belong neither to Friday nor to Saturday. This is true enough if one form of reckoning is considered. But it is evident that it might belong to Friday according to one computation,

[133] *Normae Generales,* II, 146-147.

which the person is not obliged to follow, and to Saturday according to another computation which is likewise not actually adopted.[134]

A licit separation of these two days may easily be imagined. This would occur if one crossed the International Date Line from East to West in such a way as to repeat the Friday of an ember week. The traveler in such circumstances would logically fast and abstain on Friday I because universal laws of the Church bind travelers in the same way as the faithful of the place.[135] However, there would be no obligation of fasting on Friday II since that obligation would already have been fulfilled. In that case 24 full hours would intervene between Friday and Saturday.

It is true enough that the taking of food between two ember days would violate the spirit of the law, but a similar case would exist if one took advantage on a fast day of the one full meal, which the law of fast premits, to gorge oneself and to spend the rest of the day taking such alcoholic liquors as beer and wine. Such a behavior would certainly run counter to the spirit of the law of fast, but would by no means violate the law itself. This axiom is invoked in favor of the eating of meat on Friday night without breaking one's eucharist fast for the following day, and there is no reason not to invoke it here.

A final remark must here be made. Michiels, for instance, uses canon 35 to prove that two ember days are juridically united in such a way that the same time-reckoning must necessarily be followed in the two days.[136] The argument is not *ad rem*. The continuous time mentioned in canon 35 is opposed to available

[134] Let us take the case of two consecutive ember days which occur when DST is in actual use. Sempronius reckons his Friday fast and abstinence according to DST and his Saturday fast and abstinence according to EST. Friday therefore ends at midnight DST which corresponds to 11:00 P.M. EST. Consequently, the interval of time between 11:00 P.M. and midnight EST belongs to Friday according EST (not actually followed for Friday) and to Saturday according to DST (not actually followed for Saturday).

[135] Cf. *infra*, p. 180.

[136] Cf. *supra*, p. 173.

time and is not in any way opposed to the use of different computations in two successive ember days. One is indeed certain of his ground when one states that the letter of the law does not prevent one from taking food between two successive ember days.

B—The Actual Determination of a Reckoning

Canon 33, § 1, rules that in the computation of the hours of the day the usual time of the place is to be followed; however, when the observance of certain private obligations is involved, the usual time is not imposed and one is free to use either local time, true or mean, or legal time, regional or extraordinary.

It is evident that a person cannot use more than one reckoning for the same obligation on any given day. But one might ask oneself the precise reason why one computation has in concrete cases priority over the others.

Canonists sometimes speak of the *choice* of a reckoning.[137] The word *choose* is not only an unhappy one, it is also misleading. One must bear in mind that the Code does not use the word *eligere* but the expression *"potest quis sequi"*.

The solution of the problem demands that one examine separately the negative and the positive obligations. When negative obligations are considered there is no doubt that the mere choice or intention of using this or that reckoning is of no canonical importance. One thing counts in the determination of one computation to the exclusion of the others, and that is the actual use of one of the computations. A person fulfills a negative precept by abstaining from that which is prohibited by the law. There is no necessity of having the intention of fulfilling the precept.[138] The law of abstinence, for instance, prescribes the abstention from meat on Friday, and Friday in this case is composed of 24 continuous hours reckoned according to any one of the modes mentioned in canon 33, § 1. When a person is asleep Thursday night and therefore actually abstains from meat from the first

[137] Cf. Vromant, *Normae Generales,* n. 133; *IER,* XXVII (1926), 77-80.

[138] Cf. Sabetti, *Compendium Theologiae Moralis* (3. ed., New York, 1888), n. 88. Cf. also St. Thomas, *Summa Theologica,* I-II, q. 100, art. 9.

point of time indicating a canonical midnight (DST, for instance), the obligation of abstaining from meat necessarily ends 24 hours later, whether or not the person had the intention of using that or any other reckoning.

When a person has the intention of using a certain computation for the observance of a negative precept, and even begins to observe the law according to that reckoning, he is free to change his mind provided that his action does not effectively prevent the observance of the law through 24 continuous hours. Sempronius, for instance, decides to reckon his Friday abstinence according to CST in El Paso, Tex. So he begins to abstain at 11 P. M. according to his watch Thursday night, That corresponds to midnight CST. But it so happens that unforeseen events, such as the unexpected arrival of relatives or friends, make him wish that he had not chosen CST for the reckoning of his Friday abstinence. The question is whether or not Sempronius is free to change his mind and to adopt Mountain or usual time so as to eat a chicken sandwich with his guests.

There is no doubt that Sempronius is free to do so because such a course of action violates no law and implies no contradiction. This problem is similar to that of a marriage not yet contracted. Cornelius might have the intention of marrying a certain young lady, but as long as the marriage has not been contracted, he is canonically free to change his mind and to marry some one else.

On the other hand it becomes obvious that no change of reckoning is permissible if it effectively shorten the time prescribed for the observance of the precept. Thus one cannot begin one's Friday abstinence according to Standard Time when DST is being used and cease observing the law Friday night at midnight DST. In this case 24 hours of abstinence would not have lapsed. Such would not be the case, however, if Sempronius intended to use Standard Time to determine the beginning of the Friday abstinence after he had actually abstained from meat from midnight DST.[139]

[139] Cf. Van Hove, *De Temporis Supputatione*, n. 300: "Qui efficaciter eligit unam supputationem ad implendum *unum idemque praeceptum*, illam retinere debet per totum tempus quo praeceptum urget."

One may now apply these principles to the celebration of the three Christmas Masses. One might imagine the case of a priest who had the privilege of celebrating three Masses from midnight and who reckoned his time from the earliest possible point of time for midnight, for instance, from 11 o'clock usual time in El Paso, which is midnight legal time. If it so happened that this priest inadvertently took the ablutions at the end of the first Mass, he could not then say the other two Masses on the assumption that the ablutions were taken before midnight usual time and that his fast was therefore not broken.[140] Mass is permitted at midnight precisely because it is Christmas. The first Mass is not said in neutral time. The line of conduct just referred to would imply that a certain period of time belonged at the same time to Christmas day and also to the previous day, which is absurd and contradictory.

Such would not be the case, however, if the priest had only the intention of beginning his first Mass shortly after midnight CST, but did not do so, and inadvertently took food just after the time chosen to begin Mass. He could then without the slightest doubt wait for midnight according to another reckoning and say his three Masses.

The problem of positive precepts, such as that of the recitation of the breviary, is more complicated. In these cases two distinct possibilities confront one. A person is able to determine a time-reckoning either effectively or theoretically. In the first place a cleric, for instance, effectively determines a time-reckoning, e. g. the earliest possible point of time for midnight, by reciting Prime immediately after the beginning of the first hour of the computation involved. Conversely, a mere theoretical determination exists when one does not use the privilege or latitude that any optional reckoning of time apart from the usual one might give, but simply intends to do so.

When there is an effectively determined starting point for the

[140] Cf. Van Hove, *loc. cit.;* Wernz-Vidal, *Normae Generales,* n. 248 (p. 386, note 19); Weigert, "Ieiunium Naturale und die Drei Weinachtsmessen,"—*LQS,* LXXX (1927), 335-336; Vromant, *Normae Generales,* n. 133, 2.

computation of the 24 hours, no change in one's reckoning may be made. Such a course of action would unduly give the day a duration of more than 24 hours. Let us take the case of the priest who recites Prime for Wednesday immediately after midnight DST (Tuesday entering Wednesday). If he has not recited Compline within 24 hours, that is, before midnight DST (Wednesday entering Thursday), he can no longer validly do so. The recitation of Compline after that specified time would not fulfill the precept; it would be merely an act of private devotion, irrespective of the fact that the priest was lawfully impeded or not. If one omitted Compline through downright neglect, the guilt would be irreparable in the sense that the office for that day could no longer be validly recited.

Now, other considerations enter the scene when there is no effectively determined starting point. Here again the solution of the problem is *in se* the same whether or not the element of guilt intervenes. However, *epikeia* would seem to suggest a different solution according to the existence or non existence of wilful neglect.

Father Pomponius, for instance, is in El Paso, Tex., and intends to use CST for the recitation of his office. He thinks of reciting Prime shortly after midnight CST, but for some reason or other does not do so. During the day he wilfully neglects to finish his office, foreseeing as he does that impediments in the later part of the day will prevent him from reciting the whole of his office before midnight CST. This actually occurs. If Father Pomponius is free at midnight CST he must finish his office since he is free to change the reckoning that he has chosen. If he does not change the reckoning so as to fulfill his obligation he is acting as the man who decided to pilfer a certain sum of money, and actually stole the money later on the assumption that the sin had already been committed and that the external act does not increase the guilt. In the case of Father Pomponius the precept is still fulfillable in view of canon 33, § 1 (*potest quis sequi*), since CST has not actually been used. There was merely the intention of using this time-reckoning.

Now, one can imagine the case of Father Caius who merely

intended to use CST in El Paso, Tex., but unlike Father Pomponius, was lawfully impeded in the recitation of his office. If Father Caius is free at midnight CST (midnight marking the end of the canonical day), the letter of the law demands that the office be finished just as in the case of Fr. Pomponius. The precept is still fulfillable in view of canon 33, § 1, and therefore urges.

However, it would seem unfair and inequitable to hold Fr. Caius to the complete recitation of his office. Truly, when a priest has been busy enough during the day so as to be unable to recite his office, he is entitled to a rest at midnight, whether or not it be only 11 P. M. by the usual time. But even in this supposition, if the priest actually uses the extra time to finish his office, the recitation is not merely an act of private devotion but really fulfills the precept of the Church.[141]

The consideration of this question would not be complete without a final remark. The present writer does not agree with the implicit assertion of some authors to the effect that the reckoning of time for the same obligation lasting through two or more successive days must be the same in every case for a negative obligation, and also when the obligation has not been fulfilled before the end of the day if the precept is positive.[142] Thus, according to the above mentioned contention, if Father X finished his office just before midnight according to the reckoning that he had chosen, DST for instance, he could not reckon the following day according to a different computation, Standard Time it may be supposed.

141 Canonists as a rule do not give sufficient consideration to this question, namely, to that of the determination of a time-reckoning. They simply mention the *"libera optio"* and leave one under the impression that a mere theoretical choice of one of the permissible computations necessarily excludes the others. There is no canonical justification for such a contention.—Practically the only instance mentioned with reference to the changing of a time-reckoning already adopted is that of the midnight Masses on Christmas.

142 Cf. Van Hove, *De Temporis Supputatione,* n. 300. The choice of a later reckoning for the second day is the only one considered here. No reckoning that makes the day shorter than 24 hours is permissible except in the cases explained below for travelers. Cf. *infra,* p. 181.

If this were admitted strange cases would abound. For instance, two priests choose the same reckoning for the recitation of the office. They both recite Prime shortly after midnight DST. Father Francis finishes his breviary during the day, but Father John recites Compline just before midnight. The former would then be free to choose Standard Time for the next day and would not be obliged to recite Compline before 1 A. M. DST, while the latter would necessarily have to choose the same computation as that of the previous day and would thus have to finish his office before midnight DST. Such a solution for Father John's case would certainly hold him to an obligation concerning which there is no certain proof that it actually exists.

§ 5—The Reckoning of the Day for Travelers

Thus far the reckoning of the hours of the day has been considered from a rather limited angle, that is, for persons who remain in the same place or locality. But problems of a slightly different nature arise when one goes from one place to another where a different time is used, and especially when one crosses the International Date Line (IDL) where a day is either added or suppressed.

A—When Traveling from One Zone to Another

A person boards a plane for New York at Chicago at midnight CST Thursday, which corresponds to 1 A. M. EST Friday. Incidentally, he eats meat just before taking off for New York. The question at issue is whether or not the law of abstinence ends for that person at midnight CST or at midnight EST.

The principles set down by the Code contain the key to the solution. According to canon 13, § 1, the general laws of the Church bind everywhere all those for whom they were made. Thus a New Yorker not exempt from the Friday abstinence is bound by that same law even though he be in Chicago, and vice versa, a Chicagoan is bound by the same law in New York just as well as at home. General as this law is in theory, it has nevertheless in practice an element of particularity. It binds one *hic et nunc* only through canon 33, § 1, where some form or other of local time is prescribed. The law of abstinence has therefore two elements: the general, whereby one is bound to abstain from meat on Friday, the other, particular, whereby that obligation is made to bind according to the time of the locality. It is evident that the former law (general in character) can not bind effectively if the latter law (localized in application) is not simultaneously taken into consideration. Consequently a person is affected by the general element everywhere, but by the specific or particular element only insofar as it applies to the territory in which the person actually is: *Locus regit actum.*[143]

Concrete applications may be drawn from the above principles.

1—*Fast and Abstinence Annexed to a Certain Day only*

If a person leaves Chicago at midnight Thursday CST just after eating meat and is in New York throughout Friday, then his obligation of abstaining ends at midnight EST, even though it be only 11 P. M. in Chicago where CST is used, because travelers (*peregrini*) are bound by the general laws of the Church in the same way as those who have a domicile or a quasi-domicile in the territory.[144]

In the above illustration the day becomes shorter for the traveler. Conversely, if a person went from New York to Chicago he would add an hour to his day and would theoretically have to abstain through 25 hours in view of canon 14, § 1, n. 1. In practice, however, it is reasonable to say that this person is not obliged to abstain for more than 24 hours, because he has already fulfilled the obligation of fasting through Friday which consists of 24 hours only. Van Hove states that the common teaching of the doctors maintains that travelers are not bound to observe a second time a universal law of the Church if they have already done so in some other place. He concludes by saying that this

[143] Cf. canon 14.—The traveler is free to use any legitimate reckoning of the particular locality, provided, however, that it does not trespass on the corresponding reckoning of the place he left. Thus, a person beginning his fast in Chicago accordng to CST cannot finish it in New York according to DST. Cf. Van Hove, *De Temporis Supputatione,* n. 301.

[144] Cf. canon 14, § 1, n. 1.

principle seems applicable also to the duration of a precept.[145] If the obligation of fast and abstinence lasts through more than one day, then a special difficulty arises, at least for some authors. The difficulty is that of the non-interruption of the observance of a precept which binds through two or more continuous days.

Let us examine the problem with reference to two successive ember days. If Sempronius eats meat in New York Thursday night shortly before midnight EST, and then boards a plane for Chicago, his obligation of fast and abstinence for Friday ends normally at 11 P. M. CST (midnight EST) according to the principles just enunciated. And, theoretically at least, he is not obliged to begin the observance of Saturday's precept before midnight Chicago time. Sempronius then has one hour during which he is not bound to fast and abstain.

Well, the common opinion of the canonists holds that two successive ember days impose a continuous observance of the precept of fast and abstinence through the two days. If that opinion is held, then the logical conclusion must be that Sempronius cannot interrupt his fast and abstinence at 11 P. M. CST for Friday.[146]

The present writer has already asserted that he does not consider as necessarily continuous the observance of the precept of fast and abstinence through two successive ember days. Therefore he believes that Sempronius may reckon the obligation of Friday according to EST and that of Saturday according to CST.

2—*The Eucharistic Fast*

This precept is considered from a slightly different angle. The law of abstinence binds in itself, without reference to any other act. The eucharistic fast, on the contrary, becomes binding only

[145] *De Temporis Supputatione*, n. 301, 2, note 2. Cf. Vromant, *Normae Generales*, n. 134; Wernz-Vidal, *Normae Generales*, n. 248, b, note 19; Van Hove, *De Legibus Ecclesiasticis*, n. 221.

[146] Cf. Van Hove, *De Temporis Supputatione*, n. 301, 2: "Si obligatio [ieiunii ecclesiastici et abstinentiae] per plures *dies continuos* perdurat et nullam patitur interruptionem, videtur peregrinis standum esse supputationibus loci commorationis, ita ut ieiunium et abstinentia non interrumpantur."—Cf. also Vromant, *Normae Generales*, n. 134.

in relation to the saying of Mass or to the reception of Holy Communion. Theoretically, therefore, the territory in relation to which this fast is to be reckoned is not necessarily the place in which the person actually is at the time of midnight, but the territory in which Mass is to be celebrated, or Holy Communion is to be received. Thus the priest who took a night plane from New York to Chicago with the intention of celebrating Mass at Chicago in the morning could take food until 1 A. M. New York time, which is midnight Chicago time. Conversely, the priest flying from Chicago to New York would theoretically be bound to begin his eucharistic fast at 11 P. M. in Chicago, which is midnight New York time. Many authors, however, use a more favorable interpretation for this obligation, and hold that it would then be sufficient to begin one's eucharistic fast at midnight Chicago time.[147]

Cardenas, subscribing to Pruemmer's more favorable interpretation, writes as follows:

> We believe, however, that a liberal interpretation may be followed. The common sense of the faithful prompts them to observe the midnight of the place where they are for all fasting purposes, even when they are traveling. In practice, it would not be feasible to oblige the ordinary Catholic traveler to inform himself of and to work out accurately these time problems, which offer difficulty even to the clergy.
>
> Besides, the wording "potest quis sequi loci tempus" is not so decisive as to exclude the meaning of the *locus commorationis*, that is, the actual place in which one happens to be.

[147] Such is the opinion set forth by Twomey ("The Eucharistic Fast," —*ER*, CII [1940], 414), and more or less hesitatingly by Pruemmer (*Manuale Theologiae Moralis* [4. ed., 3 vols., Friburgi Brisgoviae: Herder, 1928], III, n. 200). Van Hove (*De Temporis Supputatione*, n. 301, 1, note 1) mentions as holding the same opinion: F. Van De Loo, "Greenwich, M [idden] E [uropesche] of plaatslijke tijd,"—*Nederlandsche katholieke Stemmen*, XXIX (1929), 100-111—Van Hove himself does not explicitly reject Pruemmer's opinion, but he does not seem to adhere to it. He writes: "Est benigna quaedam interpretatio, quae in textu legis non fundatur."—*Op. cit.*, n. 301, 1. Cf. Coronata, *Institutiones*, I, n. 50, I, note 4; Beste, *Introductio*, pp. 102-104; Vromant, *Normae Generales*, n. 134.

Moreover, the whole tenor of canon 33, § 1 seems to favor a liberal interpretation in any case of a private obligation.

Hence we may conclude that one is not obliged to begin the Eucharistic fast before midnight of the actual place in which one happened to be, even though one foresees that one will say Mass or receive Holy Communion in another place where the time is different.[148]

The general conclusion to draw for the observance of the laws of fast and abstinence when one is traveling is that in no case is one obliged to fast and abstain for more than 24 hours, and that this precept ends for the traveler when it ends for the local people, even though the day be shortened. The beginning of one's eucharistic fast may be reckoned either according to the place where one actually is or according to the place where Mass is to be celebrated or Holy Communion is to be received.[149]

3—*The Recitation of the Breviary*

This obligation is the *onus diei* and must be complied with every day.[150] The precept begins to bind at midnight and ends, theoretically, 24 hours later. In practice, however, one has to use the available time at one's disposal. If one foresees an impediment for the last two or three hours of the day, the day within which the office must be recited is actually shortened. Such an impediment happens when the traveler's day is shortened by one or two hours. In that case the office must be recited within *that* day, even though it consist of only 22 hours.[151]

[148] "Midnight and Canon Law,"—*ER,* CI (1939), 410.

[149] One could imagine the case of a priest who intended to fly from New York to Chicago at 2 A.M. and consequently in view of reckoning the beginning of his eucharistic fast according to Chicago time took food just before 1 o'clock New York time. Then, for some reason or other, he misses the plane and is forced to stay in New York. There is no doubt then that the priest in question is not able to say Mass in New York because his fast has already been broken according to the local reckonings.

[150] Cf. canon 135.

[151] However, the priest who has not finished his office at midnight usual time must in theory adopt a later point of time for midnight, if there be any, to complete his office, unless he has recited Prime accord-

The same conclusions apply if the day is lengthened. Cardenas writes:

> If the day is lengthened, and one has been lawfully prevented from reciting the office for twenty-four hours, it seems that he is no longer bound to satisfy the obligation, since the time ordinarily allowed for this has passed; if, however, he was not lawfully excused, he is obliged to make use of the longer day to satisfy the precept which urges him to the last midnight.[152]

The time-element relative to the celebration of Mass and the reception of Holy Communion offers no difficulty. The first hour after noon, which marks the end of the time during which one may begin the celebration of Mass, and the point of midnight relative to the private celebration of the midnight Mass on Christmas, may be reckoned according to any one of the various computations permissible in the place where Mass is to be celebrated. This also applies to the distribution of Holy Communion.[153]

ing to an earlier reckoning in the place that he has left. If no reckoning has been effectively determined the precept urges up to the latest point of time for midnight in the place in which the traveling priest actually is. Now, if the priest's omission is due to neglect, the office must actually be recited. Conversely, if the priest has been justly impeded, the obligation does not urge *de facto*, because no one is bound to deprive himself of his legitimate sleep in order to recite the breviary.—Cf. Tanquerey, *Synopsis Theologiae Moralis et Pastoralis,* Tomus III (6. ed., Romae; Desclée, 1921), n. 1082, d. Cf. also Van Hove, *De Temporis Supputatione,* n. 301, 3.

[152] "Midnight and Canon Law,"—*ER,* CI (1939), p. 411. Cf. Van Hove (*De Temporis Supputatione,* n. 301, 3) who refers to E. Niebecker, "Die Berechnung der Tagesstunden (Zeitpunkt der Mitternacht usw.) nach dem Canon 33 des C.I.C.,"—*LQS,* LXXXIII (1930), 775.—The present writer agrees substantially with the doctrine set forth by Van Hove and Cardenas, but his reason why the priest who was lawfully impeded is not obliged to finish his office is that no one is bound to forego his legitimate sleep to recite his office. On the other hand, if the omission of the breviary was culpable, there is no question of a legitimate rest before the precept has been complied with.

[153] Cf. Van Hove, *De Temporis Supputatione,* n. 301, 4; Vromant, *Normae Generales,* n. 134.

B—*When Crossing the International Date Line*

When the traveler crosses the IDL from East to West an extra day is added. In that case the common opinion holds that there is no obligation of repeating the act which has already been fulfilled on the previous day. Thus, if Friday is repeated, on Friday II the priest traveler is no longer obliged to abstain from flesh meat and to recite the office that has already been said. On the other hand, since the possibility of offering up Mass is considered as a favor, it may be celebrated on the two days.[154]

One might now wonder whether or not the traveler in such circumstances is obliged to recite all his office on Friday I or whether it is permissible to say the whole of it on Friday II, or whether he may recite merely a part of it on Friday 1 and the remainder on Friday II.

There seems to be no doubt that the priest is free to recite the whole of his office on either day, for the simple reason that the precept will be fulfilled on the second day just as well as on the first. In fact, the priest has two days during which he may comply with this obligation. Thus, the faithful who are given a certain number of days for the fulfilling of the Paschal precept are by no means obliged to comply with it on the first or second day if they foresee no impediment for the remainder of the allotted time. It would also seem that a part of the office could be recited on Friday I and the remainder on Friday II. In so doing the priest fulfills the obligation that he has of reciting his Friday office which is the *onus diei*. It matters little whether the agent has 22 or 48 hours of available time to fulfill the precept. The unit that really counts is the day; the number of hours is merely accidental. Thus the traveler going from one zone to another must theoretically recite his office within 23 hours if the day be shortened by one hour. Conversely, when the day is lengthened by one hour, the priest may make use of this accidental extension of the day and is not obliged to finish Compline before

[154] Cf. Van Hove, *De Temporis Supputatione*, n. 301; note 2; Jombart, "La Ligne de Démarcation de Date,"—*NRT*, LV (1928), 768; Cardenas, "Midnight and Canon Law,"—*ER*, CI (1939), 413.

any one of the various points of time determinable as midnight for the locality wherein he happens to be.[155]

The same conclusions would also apply to negative precepts. Thus, the Friday abstinence could be observed on either one of the two days.

The other phase of Far East travel unravels itself when the Pacific is crossed from West to East. A day is then suppressed. In general there is little difficulty here. The suppressed day is simply non-existent, and never begins to bind. Consequently, the abstinence and the office prescribed for the suppressed Friday, for instance, may lawfully be omitted.[156]

Cardenas considers the problem of anticipating Matins and Lauds when a day is suppressed. He writes: "If one foresees on Thursday noon that the next day will be Saturday, nothing seems to prevent his anticipating Matins and Lauds for the next day, even if this day, by exception, happens to be Saturday.[157]

The celebration of the *Missa pro populo* presents a totally different problem in the present writer's opinion. Cardenas maintains that the *Missa pro populo* assigned to a day that has been suppressed has not to be said on the following day, no more than the abstinence prescribed for that day has to be observed on the following. The obligation, he holds, cannot be fulfilled on the

[155] Truly, the obligation of reciting the divine office runs through 24 continuous hours reckoned from midnight to midnight. Cf. canons 32, § 1, and 135. In practice, however, the actual time within which the breviary must be said might sometimes be contracted to two or three hours. Thus, if a priest foresees that he will be busy through the greater part of the day, and has only two hours during which he is able to recite his office, he is *de facto* bound to do so during those two hours. In the case under consideration the actual time that the traveler has to recite one day's office is 48 hours. Naturally, the present writer is considering nothing beyond the bare requirements for fulfilling one's fundamental duty of the recitation of the divine office through two calendar days.

[156] Cf. Van Hove, *De Temporis Supputatione*, n. 301, note 2; Cardenas, "Midnight and Canon Law,"—*ER*, CI (1939), 412; Jombart, "La Ligne de Démarcation de Date,"—*NRT*, LV (1928), 768.

[157] *Op. cit.*, p. 413; Cf. Jombart, *loc. cit.*

day itself, since it does not exist, nor can it be fulfilled later on, for then the day is past.[158]

Fallon justly holds a different view.[159] He argues from the totally different nature of the *Missa pro populo* as compared with the recitation of the office and the observance of the law of fast and abstinence. This difference becomes clear in the ordinary circumstances not related to the question of the omission of a day. Thus, if Father X is sick on Friday and is thereby impeded in the recitation of his breviary for that day, no one maintains that he is obliged to make up for it on the following day. On the other hand, if the pastor of a church is prevented from saying Mass on a day when the *Missa pro populo* is prescribed, he must either have the Mass said by some one else or say it himself on some other day. This is clear from the consideration of canons 339 and 446.[160]

One fact is clear, and that is that the *Missa pro populo* must be said either by the pastor or by some other priest. This obligation is not merely personal, but is also real, that is, attached to one's particular parish. No matter where the pastor is, the parish itself remains in the same geographical position and has a right to the celebration of the *Missa pro populo*.[161] The present writer unhesitatingly adheres to Fallon's opinion.

[158] *Op. cit.*, p. 413, where the same opinion is erroneously attributed to Jombart (*loc. cit.*).

[159] "Do Differences in the Calendars Affect the Obligation of the 'Missa pro Populo', "—*IER*, XXXVII (1931), 638-641.

[160] Canon 339, § 4, reads: "Episcopus Missam pro populo diebus supra indicatis per se applicare debet; si ab eius celebratione legitime impediatur, statis diebus applicet per alium; si neque id praestare possit, quamprimum vel per se ipse vel per alium applicet alia die." The same obligation applies to pastors in virtue of canon 466, § 1.

[161] Cf. Fallon, *loc. cit.*

Chapter VIII

THE RECKONING OF THE LARGER UNITS OF TIME

Article I. Introductory Remarks

The reckoning of the time-units other than the hours of the day is exclusively contained in canon 34, if one excepts the prescriptions for the *tempus utile* or the available time, which in reality considers the computation of time from a totally different angle. Available time is mentioned in canon 35.

A thorough analysis of canon 34 reveals that the reckoning of periods of more than one day is governed by two distinct and independent principles, which are, however, both applied in every concrete computation. The first of these principles states when the year and the month are to be reckoned according to the calendar, and when they are to be reckoned according to their absolute and invariable juridical value. The second general principle indicates when a period of time is to be reckoned according to the so called natural computation or according to the civil.

The natural reckoning is found either with the juridical or with the calendar computation of the month and the year. Conversely, the civil reckoning as considered in canon 34, § 3, is always according to the calendar, when calendar units are involved. In other words, continuous reckonings of months and years are always made according to the calendar, while intermittent periods follow the juridical computation.

Article II. The Reckoning According to the Calendar

Two separate problems must here be solved. The first—and the easier—considers in general when the calendar reckoning is to be used according to canon 34. The second delves into the nature of the word *calendar* and answers the question which calendar is to be used.

§ 1—When is the Calendar Reckoning to be Used?

Strange to say, the wording of the canons of Title III seems to imply that the juridical reckoning of the year and the month is the rule and that the calendar reckoning is the exception.[1]

The Code specifies when these two units are to be reckoned according to the calendar. Three instances are found:

——A The first is in canon 34, § 1, and occurs when the month and the year are designated by their proper name, as the month of February, the year 1941, or by their equivalent, as the next month, the next year, the third month, etc. In other words, the reckoning is according to the calendar when a determinate month or year is explicitly mentioned.[2]

——B The second instance is found in canon 34, § 2, where the calendar reckoning is prescribed for the natural reckoning when the time is continuous. Thus, a three months' vacation from the 28th of June, if it be reckoned continuously, will finish on the 28th of September for a total of 92 days. However, if that vacation be taken, not all at once, but on two or three occasions, for instance, for a period of six weeks from the 28th of June and for the remainder of the three months after the 15th of October, that whole vacation period is not continuous but intermittent, and in that case the juridical month of 30 days is to be used. Consequently a three months' vacation broken

[1] Thus canon 32, § 2, reads: "A month in law means a period of 30 days, and a year a period of 365 days, unless month and year are said to be taken according to the calendar."—Cicognani, *Canon Law*, p. 674.

[2] One cannot admit what Ojetti *(Normae Generales,* p. 195, 5) writes: "Suspensus *ad annum* die I ianuarii a. 1928 manet suspensus ad 365 dies, quamvis annus 1928, quia bissextus, comprehendat dies 366, nisi dictum fuerit *ad annum* 1928 vel etiam *ad totum* annum, ita ut intelligatur eum fore suspensum ad totum annum eo die incipientem."—At most, this might be true in theory, but in practice it can in no way be approved. As a matter of fact, the suspension here mentioned is *de facto* reckoned according to canon 34, § 2, and necessarily according to the calendar since the interval during which the suspension binds cannot but be continuous. And if the reckoning of the year according to the calendar has only 365 days, even when leap years are involved, the calendar reckoning is not only misleading but also meaningless. What Ojetti writes could be upheld before the Code in view of the lack of specific norms to the contrary, but it is certainly out of place at present. Wernz-Vidal *(Normae Generales,* n. 249, I) mention a theory similar to that of Ojetti.

up into two or three parts can never be more than 90 days. Thus, if a person has used the 62 days of July and August for part of his three months' vacation, he can subsequently add only 28 more days in order not to go beyond the three juridical months allowed him.[3]

——C The third instance mentioned is found in canon 34, § 3, n. 1, where it is stated that if the starting point of any period is explicitly or implicitly assigned, the months and the years are to be reckoned according to the calendar.

From all that has been said the only conclusion to draw is that the calendar reckoning is in practice the general rule, and that the only instance mentioned by the Code when the juridical computation of the month and the year is used happens to be when a period of time is reckoned from moment to moment and in addition is made up of intermittent durations of time.

§ 2—Which Calendar is to be Used?

This problem is all the more difficult for the fact that it is not squarely considered by the canonists. The few writers who treat the question of the nature of the calendar restrict themselves to the question whether or not the word *calendar* in canons 32 and 34 refers to the Gregorian calendar, as used in civil affairs today, or whether it refers to the ecclesiastical or liturgical calendar, which still indicates the days of the month according to the old Roman system of Kalends, Nones and Ides, and which considers the *sextus* and the *bissextus Kalendas Martii* as one and the same day. In other words is the leap year according to the calendar reckoning to consist of 366 days or is it invariably to consist of 365 days like the common years?

The other phase of the problem deals with the possibility of using any other calendar than the Gregorian. For instance, in

[3] The authors are not all of the same opinion concerning the nature of continuous and intermittent time *(tempus continuum, tempus intermissum)*. This problem, as such, is distinct from that which is now considered, and will be treated below. Suffice it to say here that the calendar reckoning is used when the time is continuous, and the juridical computation when the time is intermittent. Cf. *infra*, pp. 210-214.

some parts of the world the Gregorian calendar is not the one which is in common use. In that case is one to reckon one's age according to the Gregorian calendar which is not known to the faithful, or is one to use the lunar or luni-solar calendar which is the only one known? Such is the nature of the problems that are now to be examined.

—A The Problem of the Leap Month or Year

The question of the calendar as considered by the authors is whether the 24th and the 25th of February of the leap years are to be reckoned as one and the same juridical day according to the Roman system, or whether they are to be considered as two distinct days as is done in every-day life.

The classical application in this matter is the reckoning of the year of the novitiate. Thus, according to the general opinion, the civil calendar is to be followed, and if one becomes a novice at 10 A. M. February 24, 1943, that day does not count since the starting point does not coincide with midnight. Consequently, the year of the novitiate finishes the following year with the end of the day that carries the same calendar date. In other words, this novice—if he has not lost any day throughout the year of his novitiate—is free to make his profession February 25, 1944.[4]

This, however, is denied by the authors who insist on the use of the ecclesiastical calendar. They agree with the application of canon 34, §3, 3°, but their contention is that what is commonly called the 24th of February is really the *sextus Kalendas Martii,* and consequently that the end of the day that carries the same calendar date is the end of the *sextus Kalendas Martii* of the following year. But in the leap year the end of the *sextus Kalendas Martii* is the end of the second part of that day, that is, the *bissextus Kalendas Martii,* which corresponds to the 25th of February. The general conclusion of these canonists is that the person who was received as a novice on February 24, 1939. for instance, could not be allowed to make his temporary profession before February 26, 1940. This latter date could similarly have been

[4] Cf. canon 34, § 3, n. 3.

employed for the novice's temporary profession even if he had taken the religious habit February 25, 1939.

This doctrine is held after the promulgation of the Code especially by Oesterle.[5] Vermeersch also held this doctrine for a number of years after the promulgation of the Code.[6] Van Hove thus sums up the arguments in favor of this doctrine:

> *Pauci* opinantur Codicem adhibere annum ecclesiasticum in quo, anno bissextili, dies 24 et 25 februarii computantur pro uno die. Praesumendum est, aiunt, legem ecclesiasticam usurpare calendarium ecclesiasticum; insuper in annotationibus fontium remittimur ad an-

[5] Cf. *Praelectiones,* I, 19, where he writes: "Si sumitur Calendarium in sensu ecclesiastico, tunc pro Novitiatu, professione, etc., anno bissextili unus dies pro validitate actus adiici debet, prout in iure antiquo praescriptum fuit. Decisio hucusque non est facta." Cf. also the following articles by the same author: "Die Dauer des Noviziates,"—*LQS,* LXXIII (1920), 420-424; "Die Berechnung des Schalttages für das Novitiatsjahr,"—*Archiv für katholisches Kirchenrecht,* CIII (1923), 148-149. In the future this periodical will be cited in the abbreviated form *AKKR.*—Cf. also Blat, *Commentarium,* I, n. 96.

[6] Cf. especially "Supputatio Anni Novitiatus,"—*Periodica,* XVII (1928), 49*. At least in the fifth edition of the *Epitome* Vermeersch implicitly admits the common opinion. Cf. *Epitome,* I, n. 150.—Coronata holds a rather complicated view on this topic. He writes: "Si quis novitiatum ingressus sit die 28 februarii anno ordinario, et annus subsequens sit bissextilis, professionem emittere valide nequit die 29 februarii, sed die 1 martii quia ea die novitiatus coeptus est ex c. 34, § 3, n. 3; *'nam id biduum* (anni bissextilis) *pro uno die habetur'* (fr. 3. D, 4, 4; fr. 98, D. 50, 16: c, 14 X, *De V. S.,* 5, 40). Aliter vero dicendum si supputatio facienda esset iuxta n. 2 eodem canone; tunc enim annus completus esset, 27 die februarii expleta. Attamen quia annus 365 constat (c. 32, § 2) et qua talis completus est die 28 completa (c. 555, § 1, n. 2) et quia *ultima dies* eiusdem numeri (Cfr. c. 32, § 2, n. 3 [? 34, § 3, n. 3]) est dies 28, professio facta die 29 februarii validav idetur. Aliter Vermeersch, *l. c.* I, 562 (a)."—*Institutiones,* I, n. 55, 2, c, note 7.—This doctrine is to be rejected because it is obvious that the reckoning of the year of the novitiate is necessarily made according to the calendar and according to canon 34, § 3, n. 3., and this is necessary for the validity of the novitiate. Such is the tenor of a decision given by the PCI, Nov. 12, 1922—*AAS,* XIV (1922), 661. Cf. also Bouscaren, *CLD,* I, 301. And the calendar reckoning is the only one to be used in canon 34, § 3, n. 3. No other consideration is here tolerated. The question depends exclusively on whether or not the ecclesiastical calendar is to be followed.

num ecclesiasticum eiusque partes in Missali Romano; accedit praxis Sanctae Sedis, quae in multis documentis adhibet calendarium ecclesiasticum per calendas computatum. Non obstat canonem 34, § 3, n. 2, afferre exemplum "a die 15 augusti", iuxta calendarium civile, quod usu etiam veniebat in Ecclesia ante promulgationem Codicis, quando indubio in Ecclesia computatio fiebat secundum calendarium ecclesiasticum.[7]

But this doctrine cannot any longer be held.[8] In the first place the assertion that the computation of time before the Code was made according to the ecclesiastical calendar is altogether without foundation.[9] In the second place the Code itself implicitly considers the civil calendar as the one referred to. Thus, canon 32, § 2, reads: "In iure nomine mensis venit spatium 30, anni vero spatium 365 dierum, nisi mensis et annus dicantur sumendi prout sunt in calendario." If the ecclesiastical calendar were to be followed, the reference to the year as being reckoned according to the calendar could not be explained, since this time-unit would invariably consist of 365 days. Thus, the Code never mentions the week as being reckoned according to the calendar.

The fact that the Code gives the 15th of August as an example clearly indicates that the civil calendar is referred to, and not the ecclesiastical. It is to be noted, besides, that the *dies eiusdem numeri* according to the ecclesiastical calendar would present an obviously insoluble problem. Thus the Kalends, the Nones and the Ides, the *pridie Kalendas,* etc., are not numbered. Furthermore, they follow each other in irregular sequence.[10]

A conclusive proof cannot be deduced from the fact that the Holy See uses at times the old Roman system in dating documents, etc. The mode of dating has no bearing whatsoever on the matter now under consideration.[11]

[7] *De Temporis Supputatione,* n. 285.

[8] Cf. Van Hove, *loc. cit.*

[9] Cf. *supra,* pp. 73-76.

[10] Cf. *supra,* p. 14.

[11] The 24th and the 25th of February are still considered one and the same day in liturgical affairs. Thus the feast of St. Matthias is celebrated on the 24th of Feburary in common years, but on the 25th during

It is therefore safe to assert that the calendar which the framers of the Code had in mind was the Gregorian or civil calendar. The point to consider now is whether or not the Gregorian calendar must necessarily be followed every time the use of the calendar is prescribed by the Code when no explicit exception is provided for.

——B The Other Calendars

If one admits that the Gregorian calendar must always be followed unless the contrary is expressly mentioned, then insuperable difficulties from various angles would seem to arise by necessary consequence.

The writer is firmly convinced that the Gregorian calendar is not the only one that may be followed. In the first place its exclusive use is nowhere prescribed by the Code. True enough, the examples given in Title III refer to this calendar, but this is possibly due to the fact that this calendar is the most widely used. Furthermore, the framers of the Code were probably influenced by the wording of the German Code wherein the Gregorian calendar was the only one considered.

Besides the Gregorian calendar there are various other civil calendars in use today, such as the lunar calendar of the Moslems, and the luni-solar calendars used in certain parts of China for instance, even though the legal calendar there be the Gregorian.[12] There are also the liturgical and the scholastic calendars, to mention only the more common.

Theoretically at least, the usual calendar of the place is everywhere to be followed. This is the conclusion to be drawn from canon 20, inasmuch as there is no canonical prescription for the determination of the precise calendar to be adopted. Canon 20 reads:

leap years. Cf. Ojetti, *Normae Generales,* p. 195, note 9.—According to the old Roman system this feast is celebrated no sooner than on the second half of the *sextus Kalendas Martii.* All the feast days from that of St. Matthias to the end of the month are celebrated a day later in leap years. Cf. S.R.C., 9 iun. 1884—*Decreta Authentica,* Vol. III, n. 3611, ad V.

[12] Cf. *supra,* p. 7.

> If there is no explicit provision concerning some affair either in the general or the particular law, a norm of action is to be taken, unless there is question of applying a penalty, from laws given in similar cases, from the general principles of law applied with the equity proper to Canon Law, from the manner and practice of the Roman Curia, and from the common and constant teaching of approved authors.[13]

A clear example of a law given in a similar case is that found in canon 33, § 1, where the common custom of the place is prescribed in general for the reckoning of the hours of the day. Furthermore, the constant and common opinion of the doctors seems to be that the calendar actually in use in the territory is the one to be followed for the computation of time in ecclesiastical affairs.[14]

What Toso writes concerning the reckoning of the hours applies perfectly well to the calendar. These are his words:

> In primis [Legislator] statuit, horarum diei supputationem, iuxta loci communem usum fieri oportere. Cuius legis tota ratio est in congruentia, plus dicam, in necessitate legis: nam supputatio temporis a more locali, qui et iure civili semper aut introductus est aut sancitus, aliena nec concipi potest. Scilicet, poteratne Legislator populi moribus, iure civili firmatis, ignotam prorsus aut minus notam rationem supputandi statuere? [15]

It would seem strange that the Code imposed certain unknown obligations. Thus, the obligation of fasting begins at the age of twenty-one.[16] If that age is necessarily reckoned according to the Gregorian calendar, it becomes practically impossible for some people to know when this precept begins to bind. In some parts of China the faithful barely know the existence of the Gregorian calendar. let alone any practical reckoning according to that calender. In that case it would seem certain that the local calendar is to be used at least in favorable matters, for

[13] Cicognani, *Canon Law*, p. 620.

[14] Cf. *supra*, pp. 77-78.

[15] *Commentaria Minora*, I, 105. Cf. Ojetti, *Normae Generales*, p. 197.

[16] Cf. canon 1254, §2.

instance, for the reckoning of the sixtieth year, which marks the end of the obligation of fasting. The same conclusion, in theory, also holds for other obligations. However, it would seem that a person using a shorter calendar year would not be obliged to begin fasting before the end of the 21st year according to the longer Gregorian calendar. This would logically follow from the well known principle: "Odia restringi, et favores convenit ampliari."[17]

There is on clearcut exception to the use of the local calendar. It was decreed by the Holy Office concerning the reckoning of the years required for a valid marriage. It was stated that these years were to be solar years, and not lunar years. Where it was impossible to compute one's age according to solar years, an extra month was to be added to the lunar year.[18] This prescription holds even after the Code in virtue of canon 6, n. 2, since the Code has made no change with reference to the use of the calendar.[19]

One might now wonder whether the civil calendar in one of its forms is to be used to the exclusion of the liturgical and the scholastic calendars. Canon 1365 presents an interesting case. It is there stated that seminarians are to study philosophy for at least two years (*per integrum saltem biennium*) and theology for at least four full years (*saltem integro quadriennio*).[20]

No canonist, the present writer believes, maintains that the years here referred to are Gregorian years of 365 days. It is obvious that scholastic or curricular years are the ones to be followed. Such is the universal practice. The point at issue is to find a canonical explanation for this legitimate stand.

[17] Reg. 15, R.J. in VI°.

[18] S.C.S. Off. (Chan-si), 7 maii 1890—*Fontes,* n. 1122. Cf. *supra,* p. 78. It is to be noted, however, that this prescription applied only to future marriages. Marriages already contracted in which the prescribed age had been reckoned according to the lunar calendar were not to be declared null.

[19] Cf. also canon 6, n. 6. The use of the calendar as such is implicitly retained in the Code as it existed in the earlier law. Cf. Vermeersch-Creusen, *Epitome,* I, n. 76, 6*, 3.

[20] "Can. 1365.—§ 1. In philosophiam rationalem cum affinibus disciplinis alumni per integrum saltem biennium incumbant.

§ 2.—Cursus theologicus saltem integro quadriennio contineatur. . ."

There are two possibilities. The first would explain this on the ground that canon 1365 is an exception to the general norms of time-reckoning. In some instances indeed, direct reference is made of the scholastic year.[21] But this cannot be considered as a conclusive proof. No canonical evidence or certainty is to be found here. Every one knows what a scholastic year is, but its length is not defined by the Code. On the other hand it is generally known that according to canons 31 and 32 the length of the year is 365 days unless the contrary is expressly stipulated. The other explanation would not consider canons 1365 and 589, § 1, as exceptions, but merely as using a calendar reckoning of their own.

A similar difficulty arises when one is to apply the interval of one year that must lapse between the reception of the last of the minor orders and the order of subdiaconate.[22] The common and constant teaching of the doctors before the Code was that this year was not necessarily a civil year, but could be so interpreted as to mean an ecclesiastical year. Thus, one who received the last of the minor orders on Trinity Sunday, could receive the subdiaconate the following year on the corresponding day, even though only 350 days had lapsed.[23] The same opinion is maintained today by a certain number of authors.[24]

It cannot be said that this opinion is admissible even on the ground that the matter concerned is a liturgical one, and is therefore excepted from the general norms of time-reckoning as laid down in Title III.[25] It is true, indeed, that the prescriptions regulating the various intervals to be observed between the

[21] Cf. canon 976, § 2: "exeunte tertio cursus theologici anno." Cf. also canon 589, § 1.

[22] Cf. canon 978, § 2.

[23] Cf. *supra*, p. 89. The authors generally called this reckoning *moral*, but this word had a meaning slightly different from that which it usually has today.

[24] Cf. Cappello, *De Sacramentis*, II, pars III, *De Sacra Ordinatione*, n. 419: "Annus inter acolythatum et subdiaconatum potest esse sive naturalis sive etiam ecclesiasticus qui longior aut brevior est anno naturali . . . Quovis in casu annus moraliter sumendus est, non mathematice."

[25] Cf. canon 31.

reception of the different orders are closely related to liturgy, but they are nonetheless formal disciplinary laws. Even if they were considered liturgical laws, they would certainly be so in the wide acceptation of the word, and would consequently fall under the general norms of time-reckoning.[26] It must furthermore be admitted that there is no explicit exceptionary norm provided for the reckoning of these intervals.

The writer's opinion is that the general norms of time-reckoning are to be followed, but that the calendar to be used is not necessarily the Gregorian. In 1939 Pentecost Sunday fell on May 28th, but in 1940 on May 12th. Therefore the starting point for the reckoning of the year required between the last of the minor orders and the subdiaconate is either Pentecost Sunday or May 28th, and since this is a favorable matter, it would seem that one may choose either of them. When the ecclesiastical year is shorter, as is the case from Pentecost 1939 to Pentecost 1940, the ecclesiastical calendar may be followed, and vice versa, when the civil year is shorter, as will be the case for the Pentecost to Pentecost reckoning 1940-1941, in which latter year Pentecost Sunday will fall on June 1st, the civil calendar may be followed.

When the interval to be observed is indicated in months, the problem is more complicated, since the purely liturgical concept of the month does not easily lend itself to clearcut reckonings. However, Cappello maintains that the reckoning from ember day to ember day would be sufficient for a three month period.[27]

By way of summary it may be inferred that the usual calendar of the place is generally to be adhered to, that is, the usual civil calendar for the reckoning of one's age and other similar computations, and the scholastic or the liturgical calendars for matters in which these calendars form the more natural and suitable basis for the reckoning of time, precisely because they afford

[26] Cf. *supra*, pp. 111-114.

[27] *De Sacra Ordinatione*, n. 419. Cf. can. 978, § 2, where it is ruled that the priesthood cannot be conferred sooner than three months after the diaconate, and the diaconate no sooner than three months after the subdiaconate. Cf. also canon 976, § 2.

a more expedient norm and a more adaptable order in ecclesiastical administration.

Article III. The Natural and the Civil Reckoning

§ 1—The Question of the Determination of the Starting Point

The determination of the use either of the natural or of the civil reckoning is found in canon 34, §§ 2 and 3. The term "civil reckoning" is not mentioned in the canons that deal with the computation of time, but it is nonetheless obvious that it is really prescribed by the Code. Canon 32, § 1, states that the day consists of 24 hours, to be reckoned continuously from midnight to midnight. This is precisely what the civil reckoning consists in. Canon 34 § 3, is merely an application of canon 32, § 1.

Theoretically, the dividing line between the two reckonings is very clear. The civil reckoning is to be applied when the starting point is neither explicitly nor implicitly determined, while the other form is to be used when the starting point is either explicitly or implicitly determined.[28]

Unfortunately the Code does not state in what consists the determination of the starting point. Various examples, however, are given, but even if these be carefully analyzed, it is still difficult to have a clear notion of the nature of the determination of the starting point. Two examples of a non-determined starting point are given in canon 34, § 2, namely: suspension from the celebration of Mass for a month or two years, and three months' vacation in the year. On the other hand, the starting point is determined in such cases as: two months's vacation from the 15th of August, the fourteenth year of age, the year of novitiate, eight days after the vacancy of a bishopric, ten days for appeal.[29]

There seems to be no difficulty concerning the explicit assignment of the starting point, but beyond this the matter is highly controverted, especially with reference to the absence of thirty days from the novitiate.

[28] Canon 34, § 2 and § 3.
[29] Canon 34, § 3, nn. 2 and 3.

In the cases here under consideration the starting point is explicitly assigned when the Supreme Authority of the Church determines not only the duration of a period of time but also the date on which it is to begin. In other words, when a certain amount of time is granted, the legislator knows in advance the precise moment when the reckoning is to begin. That exists when the date or moment determined by the legislator has a definite place in a well ordered calendar, whether it be civil or liturgical. Such is the case when a rescript grants a two months' vacation from the fifteenth of August, or a vacation throughout the month of May for the coming year, or if any particular hour, day, week, month or year is mentioned, such as Pentecost Sunday, next week, the next month, the second year from the Pope's Coronation, provided however that the Pope be already crowned, or that the date of the coronation be determined. This in substance is admitted by the great majority of the canonists.[30]

There are very few such explicitly determined starting points illustrated in the Code. An instance is found in canon 340, § 2, where the time within which the bishops throughout the world must make their *ad limina* visits is specifically set forth.[31] In other canons special liturgical days are determined, but one day only is involved. There is no question of a starting point in the proper sense of the word.[32]

The crucial point in the matter under consideration is to find the dividing line between an implicit assignment of the starting

[30] Cf. Van Hove, *De Temporis Supputatione,* n. 304. Augustine *(Commentary,* I, 120, note 6) and Oesterle *(Praelectiones,* I, 20) maintain that the starting point is only implicitly assigned in such expressions as: Easter Sunday, Pentecost Sunday, next month, etc. It should appear obvious that these expressions do determine in an explicit manner the starting point from which time is to be reckoned. The only doubt possible would come from the lack of determination of the precise year within which must be fixed the occurrence of the Easter Sunday or of the Pentecost Sunday referred to, but such a situation or circumstance is hardly conceivable. Cf. Van Hove, *De Temporis Supputatione,* n. 304, note 1.

[31] Cf. canons 299; 300; 341; 859, § 1 and § 2; 2021, for further examples.

[32] Cf., e.g., canon 867.

point and no assignment at all. The difficulty of properly defining the nature of an implicitly determined starting point becomes apparent when one considers the divers explanations proposed by the canonists. Many definitions are merely negative. Thus, Maroto holds that there is no determinate starting point if the interval of time alone is assigned in a general way by the law, but its starting point is determined by some other distinct cause or principle, for instance by one's free will.[33] In other words, there is no implicitly assigned starting point of any kind if the agent to whom a certain period of time is conceded is free to determine the precise moment from which the period is to lapse.[34]

A similar doctrine is maintained by Oesterle and Boúúaert-Simenon. According to these authors another element is added to that of the agent's free will, namely, a casual event. There is no implicitly determined starting point if the moment from which a period of time begins to lapse depends on the agent's free will or on a casual fact.[35]

Van Hove justly remarks that these explanations are rather indeterminate. Thus, in the acquisition of a domicile, the fact of one's stay and the intention of remaining in a certain place often depend on one's free will. Yet the starting point is implicitly assigned. Likewise, disease is a casual fact, but it nevertheless implicitly determines the starting point for the reckoning of the month mentioned in canon 858, § 2.[36]

Other writers offer a positive concept in explanation of the implicit determination. According to Toso the starting point is implicitly assigned by law when it is not verbally determined,

[33] "Ex alia distincta causa vel principio, etiam, v. gr., ex libera privatorum voluntate determinatur."—*Institutiones,* I, n. 259, 2, A.

[34] Cf. Cocchi, *Commentarium,* I, n. 96, b; Chelodi-Bertagnolli, *Ius de Personis,* n. 89, b, 3; Wernz-Vidal, *Normae Generales,* n. 249, II.

[35] "Terminus nec explicite nec implicite assignatur (iuxta Boúúaert-Simenon [2. ed.], n. 192, II) quando ipsius initium lege vel superioris praecepto non determinatur, sed a libera voluntate subditi vel a facto casuali pendet, v.g., interdictum per mensem: duo in anno vacationum menses."—Oesterle, *Praelectiones,* I, 19-20.

[36] Cf. *De Temporis Supputatione,* n. 305. Cf. *infra,* p. 249.

but when it may readily be discerned from the phraseology employed in connection with the circumstances (cum verbis minime statuitur(at, verbis cum circumstantiis copulatis, facile arguitur).[37]

If these circumstances, voluntary or casual, are understood in such a way as to imply necessary starting points determined by the common law or, in other words, if they mean that such or such an act is the only one determined by law as marking the beginning of the lapse of time in such or such an institute, then one has the implicit determination of the starting point.[38]

This would then coincide with the more satisfactory explanation given by Van Hove when he writes: "*Implicita a iure fit assignatio,* quando determinatur factum, utut liberum vel casuale, a quo terminus *necessario* currere incipit, licet tempus in genere tantum determinetur. In plerisque casibus, qui per legem ordinantur, fit assignatio termini a quo implicite, si tempus statuitur continuum. Si tempus conceditur intermissum, contra, initium termini non assignatur." [39]

According to this explanation the only divergence of the implicit from the explicit assignment lies in the fact that the moment from which a certain period of time begins to lapse is not determined in advance according to the calendar, but rather as necessarily beginning from the occurrence of a specific act, whether that be voluntary or casual. Thus one's birth and one's death must be considered casual events relative to the time of their occurrence. However, one's birth necessarily determines the starting point for the reckoning of one's age. Thus also one's baptism may in this respect, that is, in relation to its date, beget important juridical consequences.[40]

Similarly certain events might be voluntary in the sense that

[37] *Commentaria Minora,* I, 109—Michiels presents the negative aspect of the same wording: "Terminus a quo . . . nec explicite nec implicite assignetur in quantum neque ex verbis neque ex circumstantiis erui possit."—*Normae Generales,* II, 152.

[38] Cf. Van Hove, *De Temporis Supputatione,* n. 305.

[39] *Loc. cit.;* Cf. Vromant, *Normae Generales,* n. 135, note 2.

[40] Cf. Wernz, *Ius Decretalium,* II, n. 113—S.C.C. *in caus. Spolet.,* 18 dec. 1627—Wernz, *loc. cit.*

the agent determines when a certain act is to be performed. Thus an aspirant to the religious life is free, provided he have the required age, to enter a novitiate one year or another. But the law stipulates that the year prescribed for the novitiate is necessarily to be reckoned from one's vesting with the religious habit, whenever that be, or from any other mode as determined in the constitutions of the religious congregation or order.[41]

As Van Hove points out, these implicitly assigned starting points are associated with the greater number, by far, of the time-periods mentioned in the Code. This occurs, for instance, in all the cases where one's age is to be reckoned;[42] in all the cases in which time begins to lapse from the notification of a certain person or the intimation of a certain thing; whenever there is involved the vacancy of a benefice,[43] or of a bishopric;[44] in the acquisition of a domicile, or of a quasi-domicile, or in the determination of one's residence for a month,[45] or of one's stay for six months in a place,[46] or of the stay of one year in Rome;[47] whenever the beginning of the prescribed time-interval is determined by one's actual residence or stay in a place, provided other secondary conditions prescribed by law be fulfilled; in the dispositions concerning the vacancy of the benefices of professed religious, when the *terminu. a quo* is determined by the profession itself;[48] in the case of the beginning of the novitiate;[49] in the renewal of the proclamation of the matrimonial banns and in the celebration of marriage after the last proclamation of these banns has been made;[50] in the observance of the prescribed time-intervals between the various orders;[51] in the case of the

[41] Cf. canon 553.

[42] Cf. the Analytical Index at the end of the Code s.v. *Aetas*.

[43] Cf. canon 155.

[44] Cf. canons 430, § 1 : 432, §1 ; Cf. also canons 161, 162, §2; 175.

[45] Canons 92; 1097, § 1, n. 2.

[46] Canon 1023.

[47] Canon 1562, § 2.

[48] Canon 584.

[49] Canon 553.

[50] Canon 1030, § 1 and § 2.

[51] Canon 978, § 2. Cf. canon 976, § 2.

ten days during which one must appeal after the promulgation of the sentence;[52] and in the casual fact, for instance, that determines the beginning of the month of illness in relation to the particular privilege in view of which the sick may under given conditions receive Holy Communion even when they are not fasting.[53]

On the other hand, if the common law does not determine the precise fact from which the time begins to lapse, there is absolutely no assignment of the starting point. Vacations for either three or two months, or for any other duration, are instances of this type of reckoning, because the law determines no fact as the necessary beginning of the vacation.[54] The beginning depends entirely on the free will of the individual. There is likewise no determined fact to mark the beginning of the time mentioned in canon 1098, nn. 1 and 2, for the person who prudently foresees that there will be no priest around for a month to assist at his marriage. That person chooses the moment from which the beginning of the month is to be reckoned and is free at his own good pleasure to change the moment from which the reckoning of the month is to begin. All the spiritual exercises prescribed by the common law also follow the natural reckoning for the reasons given above.[55]

It would not be admissible to end the consideration of this question without mentioning the most controverted problem in the Code relative to the implicit assignment of the *terminus a quo.* The point at issue is whether or not the 30 days mentioned in canon 556, § 1, which the novice may spend away from the religious house without necessarily invalidating his novitiate, have an implicitly assigned starting point. If the starting point is implicitly assigned then these days are reckoned in the civil way from midnight to midnight. Conversely, if there be no assignment the natural computation must be followed.

[52] Canon 1881.

[53] Canon 858, § 2.

[54] Cf. canons 338; 418, § 1; 465.

[55] Canons 541; 571, § 3; 595, § 1, n. 1; 1001, § 1. Cf. Van Hove, *De Temporis Supputatione*, n. 305.

Vermeersch is the foremost proponent of the civil reckoning. He argues that the beginning of the absence from the novitiate is authoritatively assigned at least if the superior grants his permission. The permission given by one in authority would be an implicit determination of the starting point. This implicit assignment of the starting point for the days of absence is similar to that of the beginning of the year of the novitiate.[56] In either case the starting point is determined by the superior, even though he have not jurisdictional authority in the full sense of the word. Consequently the civil reckoning is to be used, and if a novice were absent from 8 A. M. Tuesday to 8 A. M. Thursday, only one day of absence could be canonically computed, and that would be the 24 hours of Wednesday from midnight to midnight.[57]

Other canonists, however, maintain that the superior's permission does not change the nature of the starting point of the absence and that there is in no instance an implicit assignment of the starting point. Van Hove states that there is at least an implicitly assigned *terminus a quo* when the starting point is determined by an *act of jurisdiction,* as is done in laws, rescripts, judicial sentences, precepts proceeding from one's jurisdictional authority. The same author believes that a superior, not acting in a jurisdictional capacity, but only in virtue of the *potestas oeconomica,* is not competent to determine the starting point in the sense of canon 34. Indeed, when superiors are not exercising the power of jurisdiction they are, just as the inferiors, subject to the prescriptions of the law in those instances when

[56] Cf. canon 34, § 3, n. 3.

[57] Cf. *Epitome,* I, n. 146; "De Non Nullis Supputationibus ad Usum Religiosorum,"—*Periodica,* XVII (1928), 80*-82*. This opinion is substantially that of the following authors: Augustine, *Commentary,* III, 234-235; Haring, *Grundzüge des katholischen Kirchenrechtes* (3. ed., 2 vols., Graz: Ulrich Mosers Buchhandlung, 1924), II, 784, note 4; Pejska, *Jus Canonicum Religiosorum* (3. ed., Friburgi Brisgoviae: Herder, 1927), p. 93, B; Jansen, *Ordensrecht,* p. 158; Fanfani, *De Iure Religiosorum* (2. ed., Taurini: Marietti, 1925), n. 197, B; Raus, *Institutiones Canonicae* (2. ed., Lugduni: Vitte, 1931), p. 484, I; Coronata, *Institutiones,* I, n. 582.

time is determined by the law. And the time that a novice may spend outside the novitiate in view of canon 556, § 2, is determined by the law.[58]

The writer fully agrees with Van Hove insofar as the superior's permission does not change the nature of the starting point of the absence. It must however be stated that no jurisdictional authority other than the Supreme is able to modify the nature of the time-reckoning of the intervals found in the Code, unless that power be granted by the Code or be otherwise obtained from the proper source. Thus his distinction is irrelevant.

However, there seems to be another angle from which to examine the question of the starting point. The superior might specify the moment when the novice may leave, or the novice may leave without permission, but the moment when the days of absence begin to count is nevertheless determined by the Code itself, and that occurs when the novice physically leaves the grounds of the novitiate.[59]

Neither the superior nor the novice can modify this. This seems to be the logical conclusion of the norm given by Van Hove concerning the implicit assignment of the starting point,[60] although that author does not apply it to this case.

Thus far, therefore, it seems beyond doubt that the starting point is implicitly determined by law, and that the days are to be reckoned according to the civil computation from midnight to midnight.

The writer, however, does not admit with Vermeersch that the novice who is absent from morning to evening for any number of days, has not been absent for a single day in the canonical sense. In other words, according to Vermeersch fractions of a day are not counted and the novice must be absent through 24 consecutive hours from midnight to midnight.[61]

This seems contrary to reason. One could draw a logical con-

[58] Cf. *De Temporis Supputatione,* n. 303.

[59] Cf. canon 556, § 1. ". . . si novitius . . . e domo exierit." Cf. also § 3. ". . . licentiam manendi extra septa novitiatus . . ."

[60] Cf. *supra,* p. 203.

[61] Cf. *Epitome,* I, n. 146; n. 708, I. c.

clusion from these principles and assert that a novice could stay outside the novitiate every day of the year provided he paid a short visit to the house at least for a few minutes on 335 days (or in leap years on 336). He would then not have been absent more than 30 days.

To solve the problem one must first of all bear in mind that canon 34, § 3, n. 3, does not apply here for the simple reason that the 30 days of absence are not necessarily continuous, but may occur intermittently. In that case canon 32, § 1, is to be followed. The 30 days are merely considered as 30 distinct units of 24 hours reckoned in continuous succession from midnight to midnight.[62]

The next consideration is concerned with the practical duration of the day. The theoretical length of the day is 24 hours, and such is the case in practice also for negative obligations. Thus the law of abstinence rules that one abstain from flesh meat through the 24 hours that constitute this or that Friday, etc. Similarly, when one is given the faculty of performing this or that determinate act within a certain number of days, as is the case for elections, appeals, nominations. etc., there is no doubt as to the actual length of the days.

Positive precepts in general, however, cannot be enclosed within the 24 hour limit in such a radical and absolute way. Thus the obligation of reciting the divine office binds from midnight to midnight. However, if a priest foresees an impediment for the latter part of the day, he must finish his office before the impediment arises. In such instances the day is in some way shortened.

Likewise if one adhered too strictly to the 24 hour reckoning, the use of available time (*tempus utile*) would also lead to ludicrous consequences. Thus, if one maintains that the agent must be free to act precisely throughout the 24 hours of an available day, not a single day of an available period would ever lapse, since every one is certainly impeded for some part of every day. One must sleep, eat, drink, etc., every single day. There is also

[62] Cf. *infra*, pp. 235-237.

the case of the vacation of canons and pastors. It has been shown above that one's absence through a notable part of the day counts as a full day.

The writer would compare the case of one's absence from the novitiate to that of a pastor's vacation. Whenever a notable part of any day is spent outside the novitiate, the absence of one canonical day is to be recorded. A period of more than twelve continuous hours would seem to be a notable part of the day. Absence for a few hours is not counted in virtue of the principle that a little matter is held as nothing.[63]

This brief sketch will certainly not solve all possible doubts. It is nevertheless believed that the problem is to be solved by the application of the principles here enunciated.[64]

From what has been said to this point one may conclude that it is not always easy to determine in concrete cases when the natural reckoning is to be employed and when the civil computation is to be followed. The last word concerning the nature of the determination of the starting point from which a given interval of time is to be reckoned has not yet been said. For this reason alone some advantage would have been gained if

[63] Cf. Claeys-Boúúaert, *Selecta Capita,* p. 56.

[64] The moment to moment theory is maintained by the following canonists: Chelodi-Bertagnolli *(Ius de Personis,* n. 268, a, note 3), Voltas ("De Novitiatus Interruptione,"—*Commentarium pro Religiosis,* II [1921], 82), Michiels *(Normae Generales,* II, 150), Creusen (*Religieux et Religieuses,* n. 161), Bastien (*Directoire Canonique* [3. ed., Beyaert: Bruges, 1923], n. 111), Vromant (*Normae Generales,* n. 136), Van Hove (*De Temporis Supputatione,* n. 303), Wernz-Vidal (*De Religiosis* [Romae: apud Aedes Universitatis Gregorianae, 1933], n. 280, c) and Bakalarczyk (*De Novitiatu,* The Catholic University of America, Canon Law Studies, n. 36 [Washington: The Catholic University of America, 1927], p. 124).—From all this one thing alone is certain, and that is that an absence of one canonical day is marked against the novice who spends 24 continuous hours outside the novitiate, whether or not these 24 hours run from midnight to midnight. Nothing definite is said by these authors concerning shorter periods. One would indeed wonder whether or not repeated absences of one or two hours should be neglected or whether they are to be added so as to constitute canonical days of absence.

the natural reckoning had not been provided for by the framers of the Code.

This conclusion acquires more weight when one considers the intrinsic difficulties inherent to the natural reckoning itself. Indeed, it is not always easy to remember and to determine with precision the moment of the day from which the 24 hours are to be reckoned. No wonder then that this form of time-reckoning is little used at present in the civil jurisprudence. No mention of the natural computation is made in the German Code.[65]

§ 2—The Natural Reckoning[66]

This reckoning, as has been shown above, is to be followed when the starting point is neither explicitly nor implicitly determined.[67] Two instances of this computation are given: suspension from the celebration of Mass for a month or two years, three months' vacation during the year. It is stipulated that if the time is continuous, as in the first example, the years and the months are to be reckoned as in the calendar, but if the time is intermittent, the week is to consist of 7 days, the month of 30, and the year of 365 days.[68]

The first remark to make is concerning the meaning of the words continuous and intermittent (*continuum-intermissum*). The word continuous is a very general term and applies to any interval of time reckoned without interruption. Interruptions in time-reckoning are of two kinds. The first occurs when the agent *cannot* make use of the interval of time that is given him for a certain purpose, whether the impediment be ignorance of one's rights, or the physical impossibility of acting, etc. The determining factor of the intermission is here beyond the agent's

[65] Cf. Wernz-Vidal, *Normae Generales,* n. 242, II.

[66] This reckoning is also called the moment to moment reckoning, or the hour to hour computation. Cf. Maroto, *Institutiones,* I, n. 257, 3, note 1; S.C. de Relig., *Cum propositae,* 3 maii 1914—*AAS,* VI (1914), 229.

[67] This concept of the natural reckoning is new in Canon Law. Before the Code this computation was more often used when the starting point was in some way determined. Cf. *supra,* p. 79.

[68] Canon 34, § 2.

control. Interruptions of this kind are considered in Canon Law only in certain instances. The time-intervals susceptible of such interruptions are called *tempus utile* or available time. These are considered in canon 35. Continuous time here means an interval which lapses whether or not the agent is able to use it.

The other form of interruption depends on the *free will* of the agent, and is considered in canon 34, § 2. Two elements are here to be considered. There is first of all the legal element which concedes to the agent the right to break up a time-interval in two, three or more sections. An instance of this is found in canon 338, § 2, where it is decreed that residential bishops may take a three months' vacation every year and that these months may be either continuous or intermittent. Here the second element intervenes. The bishop decides whether his vacation will be taken on one and the same continuous occasion, or on two or three distinct and separate occasions.

When a period of time is actually broken in two or three parts by the agent's will, this time-interval is then said to be intermittent. Conversely, if the interval is taken without interruption, it is then looked upon as *tempus continuum de facto,* and is opposed to *tempus continuum de iure,* which refers to an interval which must necessarily be reckoned without interruption, such as is the time in one's age, etc.[69]

That seems to be the more logical interpretation of *tempus*

[69] Cf. Michiels, *Normae Generales,* II, 153, note 1.—Coronata (*Institutiones,* I, n. 56, 1) mentions the difference that exists between intermittent time in general and between intermittent time in some special cases. In general the agent is free to break up a time-interval in as many sections as it has days. These interruptions do not affect the essence of time. Thus a pastor may interrupt his vacation four or five times, but he nevertheless has a right to 60 full vacation days. The time units are displaced, but not modified. A different case of interruption is found in canon 556, § 2, where time-units are actually dropped in some instances. Thus, if a novice spends fewer than 15 days outside the novitiate during the prescribed year, this period is nevertheless counted as a full year and there is no obligation of supplying the days that were spent outside the novitiate grounds. The days of absence are to be added to the regular year only if they be more than 15 and not more than 30.

continuum de iure and *tempus continuum de facto.* Van Hove, however, considers the problem from a slightly different point of view. He writes that the Code has in view only those time-intervals that the law prescribes as continuous or intermittent. If a time-interval is *de iure* prescribed *only* as intermittent, then the time must necessarily lapse intermittently. Conversely, when a time-interval is prescribed *also* as intermittent, that same time-interval is *de iure* conceded as continuous if the agent reckons it as such.[70]

According to Van Hove the three months mentioned in canon 338, § 2, are *de iure* either continuous or intermittent. In other words, a bishop observes the law whether his vacation be *de facto* continuous or intermittent. This distinction does not seem to serve any determinate purpose. Indeed, all the time-intervals that are here considered are legal intervals in the sense that any extension or use of time that is opposed to the canons is illegal and deserves no attention here. This theory, furthermore, does not allow one to distinguish between an interval that must be reckoned as continuous and the periods that are not necessarily so computed.

The question is merely theoretical, but it paves the way for a practical one. The point at issue is to know whether or not continuous time, which is such merely *de facto,* as in a vacation, is to be reckoned according to calendar, or necessarily according to the juridical month and year of 30 and 365 days respectively.

Ojetti strongly insists that a vacation of three months, for instance, begun on August 15 would not run continuously to November 15 but only to the thirteenth.[71]

[70] *De Temporis Supputatione,* n. 302, note 1: "Codex agit tantum de temporibus in iure statutis, ut continuis vel intermissis. Si tempus de iure statuitur *tantum* intermissum, intermisse tantum servari potest, sed statuendo tempus *etiam* intermissum, simul de iure concedit tempus ut contiuum arbitrio utentium."—The present writer knows of no instance in the Code of a *tempus de iure tantum intermissum.*

[71] Cf. *Normae Generales,* p. 202, 2*: He writes: "Exinde apparet non esse verum illud, quod advertit CHELODI, *Ius de personis,* pag. 147, not. 5, dicens: 'Canonicus, qui habet tres menses vacationum (c. 418, § 1), si in villam discedit die 1 iulii et sine interruptione rusticatur, die 1 octobris,

Michiels rejects this opinion with forceful language. He judiciously asserts that the law determines whether or not any given time-interval may be reckoned intermittently, but that it depends on the will of the agent to determine whether the time-interval involved is to be *de facto* reckoned continuously or intermittently.[72]

The first thing to do when one is faced with a theory that puts a restraint on one's liberty is to inquire whether or not it be canonically justified. Ojetti gives no proof for his assertion. The only thing that canon 34, § 2, does is to give an example of continuous time, and that is the suspension from the celebration of Mass for one month, or two years, etc. It is to be noted that the Code uses the words *"ut in allato primo exemplo"* when speaking of continuous time, and does not mention the second example when referring to intermittent time. The stressing of the first example is explained by the fact that

eadem hora qua abiit, redire tenetur et fruitus est 92 diebus: si per annum variis temporibus absens fuit, summa omnium dierum et horarum 90 dies superare nequit.' Aequivocum inde ortum est, quod CHELODI putat *continuum et intermissum* hic sumendum esse subiective, idest in linea exsecutionis, dum e contra obiective sumendum est et in linea obligationis. Aliis verbis, continuum et intermissum dicitur respectu habito ad legem, de qua agitur, non ad exsecutionem legis, quae ab obligato fiat. Quod patet enim ex eo, quod exemplum allatum ab ipso Codice de tempore intermisso respicit etiam casum, quo lex potest etiam non intermisse servari."—*Op. cit.*, p. 202, note 1.

[72] "Haec sententia videtur omnino rejicienda. Sane, ab ipsa lege pendet, num tempus in casu determinato dici debeat indivisibile seu continuum, sine interruptione computandum, an potius divisibile seu intermissioni obnoxium; in hoc altero casu tamen ab exsecutore pendet, num de facto illud intermittat necne. Jamvero, nullatenus videtur dubium, 'tempus intermissum' in hoc canone significare tempus *de facto* (legitime sane) intermissum; non videtur enim quamnam ob rationem, in casu temporis intermissioni obnoxii sed de facto non intermissi, statueretur recursus ad supputationem temporis *fictitiam,* cum tunc sine ulla difficultate recurri possit ad supputationem *realem,* calendario consentaneam? De cetero, ad exemplum allatum quod attinet, notetur, erronee affirmari quod vacatio canonici in casu cessat "ineunte die 13"; tempus enim, in casu, computandum est, non de die in diem, sed de momento ad momentum."—*Normae Generales,* II, 153, note 1.

the time of a suspension is always continuous, while that of a vacation may be either continuous or intermittent. Its juridical computation therefore depends on whether it is actually taken continuously or intermittently.

If the word *continuum* in canon 34, § 2, is to mean *continuum de iure tantum,* then, to be sincere, one must admit that the word intermissum, which stands in contradistinction to *continuum,* must also mean *intermissum de iure tantum.* A time-interval of this *tempus intermissum de iure tantum* would necessarily have to be reckoned intermittently. To say the least, it is hard to conceive the possibility of such a mode of time-reckoning. The present writer knows of no concrete example in the Code of this *tempus intermissum de iure tantum.*

In fine, the more common opinion of the canonists on this point is that a three months' vacation, provided it be *de facto* taken without intermission, is not limited to 90 days, but is to be reckoned from moment to moment according to the calendar.[73]

The last remark to make concerning the natural reckoning is applied to the first example given in canon 34, § 2, of an interval of time in which the starting point is neither explicitly nor implicitly assigned, and that example is a suspension from the celebration of Mass for a month or two years. This example does not seem to agree with the general norms of interpretation that have been given above concerning the nature of the determination of the starting point.

Suspensions are of two distinct kinds. Some are censures, but these need not be discussed here since their duration is not determined by any preordained time-interval, but fundamentally depends on the party's lack of amendment. When the guilty party

[73] Cf. Van Hove, *De Temporis Supputatione,* n. 302, note 1, who arrives at the same conclusion as Michiels, but from a slightly different mode of argumentation; Chelodi-Bertagnolli, *Ius de Personis,* p. 159, note 4.—Vermeersch, however, seems to admit Ojetti's theory when he writes: "Sumuntur 30 vel 365 dies, quando tempus designatur *intermissum,* ut cum canonico beneficiato absentia trium mensium permittitur."—*Epitome,* I, 149, 2.

recedes from his contumacy the ecclesiastical superior is then bound to absolve him from the censure.[74]

Suspensions are in all other instances vindictive penalties. In this case their duration does not depend on the cessation of one's obstinacy. The penalty must invariably continue through the prescribed time unless a dispensation be given by the legitimate superior.[75]

But suspensions as vindictive penalties can begin only in two determinate ways. The first happens when a suspension is incurred *ipso facto*. In other words the penalty is incurred without the intervention of a condemnatory sentence and begins to bind from the moment of the perpetration of the misdeed to which the law has annexed the penalty in question. An example of this is found in canon 2373.[76]

According to this canon a bishop incurs *ipso facto* a suspension of one year from the time of the conferring of orders if he ordains any one illicitly in one of the four ways specified. The starting point for this year of suspension is obviously determined. It is the illicit ordination itself. An instance of this would be the ordination of a candidate who is not a subject of the minister of ordination, the while the ordaining prelate is not in possession of the proper dimissorial letters. No other crime or misdemeanor than the four mentioned in canon 2373 could

[74] Cf. canons 2255, § 2, and 2248, §§ 2 and 3.

[75] Cf. canon 2289.

[76] "In suspensionem per annum ab ordinum collatione Sedi Apostolicae reservatam ipso facto incurrunt:

1.° Qui contra praescriptum can. 995, alienum subditum sine Ordinarii proprii litteris dimissoriis ordinaverint;

2.° Qui subditum proprium, qui alibi tanto tempore moratus sit ut canonicum impedimentum contrahere ibi potuerit, ordinaverint contra praescriptum can. 993, n. 4, 994;

3.° Qui aliquem ad ordines maiores sine titulo canonico promoverint contra praescriptum can. 974, § 1, n. 7;

4.° Qui, salvo legitimo privilegio, religiosum, ad familiam pertinentem quae sit extra territorium ipsius ordinantis, promoverint etiam cum litteris dimissorialibus proprii Superioris, nisi legitime probatum fuerit aliquem e casibus occurrere, de quibus in can. 966."—Cf. also canon 2410.

mark the beginning of the particular penalty here mentioned. The ordaining bishop is naturally physically free to ordain or not to ordain any one without dimissorial letters, but once the ordination has been conferred by him in this illicit manner, the penalty is necessarily incurred without the intervention of the bishop's free will.

In many other instances suspensions are not incurred *ipso facto,* but only through a condemnatory sentence. There are many such cases mentioned in Book V of the Code, but in no case is the duration of the suspension prescribed. It is left to the discretion of the ordinary, who determines the duration of the penalty according to the gravity of the offense and in line with other pertinent considerations.

When the duration of a time-interval is not determined by the common law the norms for the reckoning of time found in the Code are not necessarily to be observed. They would have to be followed, however, if the Holy See intervened in an administrative way and decreed that one was suspended for a certain amount of time. In this case also there would be a determinate starting point for the reckoning of the time-interval. This determinate starting point would be the official notification of the suspension.

There is absolutely no difference as to the nature of the determination of the starting point with reference to the ten days that one may use to appeal from an ordinary court sentence, and with regard to the two months during which one would be suspended. In spite of that the ten days given to appeal are mentioned in canon 34, § 3, n. 3, as having an implicitly determined starting point, while a suspension of two years is given as an example of an interval of time without a determined starting point. It is difficult to find a satisfactory explanation why the time for suspensions is not considered by the Code as having an implicitly assigned starting point.

In the present writer's opinion a better example for the natural reckoning would have been the time-interval of some spiritual exercise prescribed by the Code, as the retreat of at least eight full days before a postulant is admitted to the novitiate.[77] In this case there is no determinate act set down by the law as marking the beginning of a retreat. On the other hand the beginning of a suspension comes as a direct result either of the perpetration of a certain misdeed or of the official notification of one's suspension by a juridical sentence of some kind.

One may conclude by saying that a suspension incurred *ipso facto* should theoretically follow the civil reckoning. Thus, if a suspension of one year were incurred on the second of April, 1941, the penalty would not cease before midnight April 2-April 3, 1942. In practice, however, it would be permissible to use the moment to moment reckoning in virtue of the example given in canon 34, § 2. It seems strange that the authors do not mention this at least apparent contradiction.

§ 3—The Civil Reckoning

According to this form of reckoning the period of time conceivable as a day consists solely and exclusively of 24 continuous hours from midnight to midnight. Parts of a day are not considered here. As already mentioned, this form of computation is the one actually used when the starting point is at least implicitly assigned.

Canon 34, § 3, which, so to say, lays the cornerstone of the civil reckoning, contains four distinct prescriptions that are to be applied in the reckoning of those time-intervals which consist of one or several months or years, one or several weeks, or several days, and whose starting point is explicitly or implicitly specified.

The first of these prescriptions determines the duration of the month and the year by stating that these units are to be reckoned according to the calendar. The week is not mentioned since it never varies. The second prescription, of greater importance than the first, deals with the determination of the first day that is to be counted. The third solves the problem of the reckoning according to the calendar when the month does not

[77] Canon 541.

have a day of the same date, or when the month in which falls the *terminus ad quem* is shorter than the month in which falls the *terminus a quo*. The fourth provides a special computation for acts of the same kind that are to be repeated.

——A Determination of the First Juridical Day [78]

The Code considers two possibilities as to the determination of the first juridical day of any enumeration. It considers the starting point either as coinciding with the beginning of the calendar day (midnight), or as not coinciding with it. If the starting point coincides with the beginning of the day, as two months' vacation from the fifteenth of August, the first day is counted and the interval of time ends with the beginning of the last day bearing the same numerical date.[79] In other words, August 15 is counted and the two months end at midnight October 14th-October 15th.

Now, if the starting point does not coincide with the beginning of the day, and such is the case when one's age is reckoned, or when one computes the time for the year of the novitiate, or the eight days after the vacancy of a bishopric, or the ten days given to appeal, etc., the first day, that is, the day on which the allotted time begins its course, is not counted, and the designated time reaches its close only when the last day of the period or interval in question has been ended.[80] For instance, if one is born at 1 A. M. January 20, 1940, the precise moment of birth does not coincide with midnight. Consequently the first day is not counted, and that person will attain the juridical age of 14 years at midnight January 20-January 21, 1954. As one sees. the 23 hours that lapse on January 20, 1940, do not count. If by rightful assumption it could be determined that the person had been born just one hour sooner then such a person would have gained a full day in the reckoning of his age.

A few expressions seem to need explanation. The first is the

[78] The first prescription of canon 34, § 3, relative to the calendar reckoning of the month and the year has already been considered. Cf. *supra*, p. 191.

[79] Canon 34, §3, n. 2. [80] Canon 34, § 3, n. 3.

question of the beginning of the day. Augustine, contrary to the common and universal teaching of the authors, states that the beginning of the day is taken in a moral way so as to mean something like 9 o'clock in the morning.[81]

It is obvious that the beginning of the day referred to in this paragraph is the beginning of the canonical day mentioned in canon 32, § 1, as consisting of 24 hours that are reckoned continuously from midnight to midnight. There is no exception in the law to warrant a different interpretation here.[82]

The second remark deals with the coincidence of the starting point with the beginning of the day. There are two kinds of coincidence: the juridical and the physical. An example of the former happens when one is given two months' vacation from August 15th. This vacation begins, or rather the time of the vacation begins to lapse at midnight August 14-August 15, whether or not the vacationer actually takes advantage of it. In other words, this coincidence is prearranged by the law itself and does not depend on a physical fact or casual event.[83]

The other type of coincidence occurs when an event of some kind, which the law in particular determines as the starting point for the reckoning of a certain interval of time, takes place precisely at midnight. In the case of the juridical coincidence the starting point determined by law was midnight itself. In this physical coincidence, conversely, the starting point is a casual event which might or might not coincide with midnight. It is safe to say that in the ordinary run of human affairs the

[81] *Commentary,* I, 122.

[82] Cf. Van Hove, *De Temporis Supputatione,* n. 313.

[83] An instance of a time-interval whose starting point coincides with midnight is found in canon 9 where it is ruled that the laws promulgated in the *Acta Apostolicae Sedis* do not become binding until three months have elapsed after the date affixed to the number of the *Acta* in which they appear. The starting point for the reckoning of the three months is not the moment of the printing of the Acta, but the more or less fictitious date that appears on the various numbers of this publication. This date coincides with a calendar day reckoned from midnight to midnight.—Cf. Vromant, *Normae Generales,* n. 137. For the reckoning of time in rescripts Cf. *op. cit.,* n. 145.

occurrence of the physical coincidence is very rare. It might happen in the case of a novice leaving the novitiate, in the case of a sudden death, and other similar instantaneous actions. On the other hand, some acts cannot be performed in an instant and cannot therefore be said to coincide with midnight. Acts of such a nature are exemplified, for instance, in a child's birth, or in the ceremony attendant upon one's entrance in the novitiate.

This might help to explain the fact that the Code mentioned these and other casual events as not coinciding with midnight. It is because they usually do not take place precisely at midnight. If by way of deviation from the normal course of happenings they did in reality coincide with midnight there would be no reason to exclude that day from the computation.[84]

A final remark on this particular topic must be made about the "ultimus dies eiusdem numeri". This expression must be taken in a wide sense and not in the strict acceptation of the word. If it were so taken the recurrence of a day with an identical numerical date would apply only when the interval of time to be reckoned was expressed in years and months to the exclusion of the other units. On the other hand, canon 34, § 3, must also provide a norm for the civil reckoning of other periods, such as a week, 10 or 15 days, etc. Van Hove writes:

> Si autem tempus conceditur non per mensem vel annum, sed per hebdomadas vel dies, dies eiusdem numeri recurrit *in hebdomada* quando recurrit eadem feriatio et terminus computatur de Dominica in Dominicam, de feria II ad feriam II sequentis hebdomadae et ita porro; in *octiduo* vel *decendio* et aliis similibus deest dies eiusdem numeri, sed dici potest diem octavum vel decimum recurrere, quando recurrit dies octavus vel decimus post diem initialem temporis. In decendio a 1 ianuarii, dies eiusdem numeri recurrit 11 ianuarii.[85]

[84] Cf. Van Hove, *De Temporis Supputatione,* n. 313.

[85] *De Temporis Supputatione,* n. 309. Cf. Vermeersch-Creusen, *Epitome,* I, n. 150. A remark might here be made concerning the reckoning of an interval of one day in the light of canon 34, § 3, n. 3. When an interval of one single day has no determinate starting point, the time is then reckoned according to canon 34, § 2, that is, from moment to moment.

In like manner the reckoning of a single day is without difficulty if a

——B Reckoning When A Day of Identical Date is Lacking

Canon 34, § 3, n. 4, reads: "And if the month has no day of the same number [no day with an identical date], e.g., one month from January 30, then, duly considering diverse cases, the term expires either at the beginning or the end of the last day of such month." [86]

There seems to be some difficulty in the interpretation of canon 34, § 3, n. 4, in consonance with the other numbers of canon 34, § 3. Chelodi sees at least an apparent contradiction within § 3 of canon 34. He states that a month from April 30 according to the calendar reckoning would be May 31, but that a month according to the prescriptions of n. 2, and n. 3, of canon 34, § 3, would expire either at the beginning or the end of May 30. He adds that § 3 of canon 34 is taken nearly verbatim from No. 187

calendrical day is specified (e.g. Easter Sunday, July 12, etc.,) or when the starting point for the reckoning of the day involved coincides with midnight. However, there is no provision in Title III for the reckoning of a single day with a determinate starting point which does not coincide with midnight. One might imagine a certain rescript requiring the positing of a definite act within one day after the reception of said rescript. The receipt of the rescript constitutes a determinate starting point and thereby excludes the moment to moment reckoning mentioned in canon 34, § 2. The day mentioned in the rescript is theoretically to consist of 24 hours reckoned from midnight to midnight according to canon 32, § 1, but the point at issue is to determine whether that day is the day on which the rescript is received or the following day. If canon 34, § 3, mentioned the reckoning of a single day along with the other time-intervals there would be no difficulty. In that case, if the rescript were received at 8 A. M. July 16, that day would not count and the day would expire at midnight July 17-July 18 according to canon 34, § 3, n. 3.

Canonical equity and general legal principles can easily supply the text of the law. Canon 20 may here be invoked. There is no express prescription of the law to reckon the single day in the case now under consideration, but there is a prescription of the law for the reckoning of time in similar instances when intervals of more than one day are considered (canon 34, § 3, n. 3.). Furthermore incomplete days are not generally counted as *days* in legal affairs. The present writer's conclusion is that canon 34, § 3, n. 3, applies also to the reckoning of a single day. Such is in substance the opinion of Van Hove *(De Temporis Supputatione,* n. 313 [p. 274, note 2]).

[86] Cicognani, *Canon Law,* p. 685.

and No. 188 of the German Code, with the exception of n. 1, of canon 34, § 3, which is lacking in the German Code.[87]

Coronata finds a solution for the problem presented by Chelodi. He is of the opinion that the apparent contradiction here referred to is obviated if the prescription of n. 1, of canon 34, §3, is followed in every instance. This n. 1, indeed, is in consonance with the rules given in § 1, and n. 4, of § 3, of canon 34. The prescriptions contained in n. 2, and n. 3, of canon 34, § 3, are to be applied only when the *dies eiusdem numeri* [day with an identical date] really occurs in the second month, but this is not always the case. When this *dies eiusdem numeri* does not occur in the second month, then the prescriptions of n. 2, and n. 3, are to be modified in the light of n. 1, and n. 4.[88]

The present writer is possibly mistaken, but he fails to see any difficulty in the simultaneous use of all these prescriptions concerning the calendar. Canon 34, § 3, n. 1, simply states the general rule that the month and the year are to be reckoned according to the calendar. This is a very general statement and needs the determinations that are found in the following numbers of the same paragraph. Thus nn. 2 and 3 simply determine which day is to be considered the first one of the enumeration from the juridical point of view. These two numbers must always be applied, since without them there would be no norm for the determination of the first day.

The root of the difficulty is that Chelodi tried to imagine a reckoning according to the calendar which did not conform with the computation prescribed in nn. 2 and 3. Thus Chelodi maintains that a month from April 30 according to the calendar would end May 31, while according to nn. 2 and 3, it would end either at the beginning or at the end of May 30. The first part of the argument does not hold. Indeed, the words "from April 30th" must be so interpreted that the starting point either does or does not coincide with the beginning of the day (April 30). If it does coincide, the month will have lapsed with the

[87] Chelodi-Bertagnolli, *Ius de Personis*, n. 89, b, 4, note 5.

[88] *Institutiones*, I, n. 55 (p. 52, note 9).

beginning of May 30, i.e., with midnight May 29-May 30. If the starting point does not coincide with the beginning of the day, that day does not count then and the interval of time will have lapsed at the end of the day which bears the same numerical date, that is, at midnight May 30-May 31. In no case in the present example does the month end on May 31.

N. 4 of canon 34, §3, goes a bit farther and considers those extraordinary instances when a day with the same numerical date does not occur in the second or last month. February of the common years is three days shorter than January. For instance, one month from January 31st will expire either at midnight February 27-February 28 or at midnight February 28-March 1. A vacation of one month starting from January 31 would have a starting point coinciding with the beginning of that day. That first day (January 31) would then be counted and one might quite naturally expect that the month would finish with the beginning of February 31. But, since the month of February has no day identical in date with January 31, the month ends with the actual last day of February, and indeed just as effectively as if a real day of the same date had existed. In the present case the designated month reaches its close just as soon as the last day of February has begun. The end of the designated month is coincident with the beginning of February 28.

On the other hand, the month starting from the vacancy of a bishopric would have a starting point that does not coincide with the beginning of the day on which the bishopric becomes vacant, and, therefore, a month from January 31 in that case would finish in theory with the end of February 31, or, in practice, with the end of the actual last day of February, which is the 28th in common years.

It might seem more reasonable to assume that the month starting from the 31, or even the 30, of January should invariably close with midnight Feburay 28-March 1, and not with midnight February 27-February 28, as is sometimes the case. According to the prescriptions of the German Code a month from January 31 invariably expires at midnight February 28-March 1, of the common years, and midnight February 29-March 1, of the leap years.[89] The framers of our Code, however took a different stand and wanted the last day of the shorter month to be reckoned similarly to the *dies eiusdem numeri.*

——C The Repetition of Acts of the Same Nature

A special reckoning is provided for actions of the same nature which are to be repeated. The disposition of the law is as follows: "In the case of periods appointed for the renewal of acts of the same kind, e.g., the three years' interval between the taking of temporary and perpetual vows, three years or any other period of time between elections, etc., the time expires on the same day of the month on which it began, but the new act may take place throughout the entire recurring day." [90]

If one made one's temporary profession at 10 A. M. January 1, 1940, the three prescribed years, according to the general law found in canon 34, § 3, n. 3, would not expire before midnight January 1-January 2, 1943, and the perpetual profession could not be taken before January 2 of that year. But, in virtue of n. 5 the perpetual profession may be made at any hour of the day January 1. Nearly all the authors agree on what is the practical meaning of this n. 5, but its theoretical interpretation follows a wide range of opinions. The only divergence of a practical importance is the question whether or not an act must be actually repeated in order to be reckoned according to this special computation, or whether it be sufficient that the act be of such a nature that generally it is repeated.

In spite of the diversity of opinion as to some of the points, no one seems to disagree with the general doctrine according to which the acts for which this special reckoning is prescribed need not be of the same species, but may also be merely of the same genus. Thus a temporary and a perpetual religious profession are similar acts of the same genus but not of the same species, as would be two temporary professions. All these

[89] Cf. *supra,* p. 106.

[90] Canon 34, § 3, n. 5—Cicognani, *Canon Law* p. 685.

various acts, therefore, enjoy the privilege of this special computation.[91]

The theoretical aspect of the controversy centers about the interpretation of the words *"recurrente eodem die"*. According to a certain number of authors the prescriptions of n. 5 are altogether exceptional and do not in any way refer to nn. 2 and 3 of the same paragraph. The day on which an act is posited counts, contrary to the dispositions of canon 34, § 3, n. 3, so that the words *"recurrente eodem die"* invariably refer to the beginning of the day of the same date. Thus, a temporary profession for one year made on July 1, 1940, would expire at midnight June 30-July 1, 1940, or technically with the beginning of July 1. This is a radical exception to the general norms given in canon 34, § 3, nn. 2 and 3. According to canon 34, § 3, n. 3, the interval would expire at midnight July 1-July 2. However, in spite of the fact that the interval has completely lapsed, one is nevertheless given that full day (July 1) for the renewal of one's profession.[92]

This derogation to the general rule is not absolute but merely relative. It applies, writes Van Hove, only in the case of acts that must be effectively renewed, such as one temporary profession after another. It does not hold, however, for acts that are not necessarily renewed, but whose renewal depends on the free will of the agent. Such is the case with the perpetual profession which is not obligatory. When the three years of the temporary vows have expired one is free to leave the religion.[93]

This interpretation does not seem to find a solution for all the possibilities and seems furthermore in opposition to the text of the law. Thus, n. 5 itself gives as an example the case

[91] Cf. Van Hove, *De Temporis Supputatione,* n. 314.

[92] This theory is maintained by Ojetti *(Normae Generales,* p. 204), Oesterle *(Praelectiones,* I, 21) and Cicognani (*Canon Law,* p. 690), the latter of whom teaches, however, that in the renewal of a religious profession the time of the first profession is prolonged and extended, as continuous, so as to end with the new profession. Cf. Van Hove, *De Temporis Supputatione,* n. 315, 1.

[93] Cf. *op. cit.,* n. 315, 1.

of the perpetual profession that is to be made after the temporary. There is no obligation imposed upon any one to make a perpetual profession once the three years of the temporary vows have lapsed.

A second group of canonists maintain that there is absolutely no exception to the general rules of time-reckoning as expounded in nn. 2 and 3. If an act that is to be renewed does not coincide with midnight, the day on which it occurs does not count, and the interval therefore expires with the end of the final day of the same date similarly as in n. 3. On the other hand, if that act coincides with midnight, the day counts, and the time expires with the beginning of the day bearing the same date. The only difference consists in this that acts which are to be renewed may be effectively renewed at any time throughout that day. If the act is not renewed, the time-interval is reckoned according to the general rule and expires with the beginning or with the end of the final day according to whether the starting point did or did not coincide with the beginning of the initial day.[94]

There is also a third interpretation of n. 5 according to which the general norms are not to be adhered to. The words *"recurrente eodem die"* are to be understood in a very broad way as referring to that day in general. The interval of time is regarded as actually expiring when the new act is posited. If the act is not renewed, the interval expires at the person's discretion, when he so chooses. Thus, if one decided not to take the perpetual profession, and had made his temporary profession on July 1, he would be free to leave after midnight of June 30-July 1 three years later. The matter is thus explained by Van Hove:

> *Per totum ultimum diem temporis* terminum ad quem currere alii tenent et proinde tempus finitur ultimo re-

[94] This theory is expounded by the following authors: Maroto *(Institutiones,* I, n. 259, d), Lacau (*De Tempore,* n. 53), Cance (*Le Code de Droit* Canonique, I, n. 74, b). Hilling (*Die allgemeinen Normen des C. i. c.,* p. 154), F. A. Schweigman ("De Geldingsduur eener Tijdelijke Professie,"—*Nederlandsche katholieke Stemmen,* XXIX [1929], 148-151, Michiels (*Normae Generales,* II, 156.)

> currente die, quavis hora, arbitrio agentis electa, sed per integrum diem, usque ad mediam noctem, actus potest renovari. "Recurrente die", id est adveniente die, iam a media nocte sed etiam usque ad mediam noctem sequentem, seu per integrum diem, actus poni potest, quo prius tempus finiatur et novum, si locus sit, incipiat. Inde si actus renovetur, terminus ad quem finitur per renovationem actus. Si non renovetur, terminus finitur expleto die: agitur enim de actibus quorum initium non coincidit cum initio diei. Si denique actus potest renovari vel non, puta professionem perpetuam post temporariam, tempus finietur hora electa ab illo qui non intendit actum renovare. Derogatur regulis supputationis temporis tantum, quatenus dies termini ad quem incompletus pro completo habetur.[95]

This opinion is strongly maintained by Vermeersch on the ground that canon 34, § 3, n. 5, is not an exception to numbers 2 and 3 of canon 34, § 3, but is to be used instead of them when the renewal of acts of the same kind is involved. In other words, the professed religious is free to determine the precise moment of the day when the vows cease to bind. Thus, if one made one's temporary profession July 1, 1940, and decided not to make the perpetual profession, one would then be free to leave the convent at any time within the 24 hours that follow upon the moment of midnight June 30-July 1, 1943.

According to the interpretation proposed by Vermeersch canon 34, § 3, n. 5, would be a special instance where the old rule is applied: *dies coeptus habetur pro completo.*[96]

[95] *De Temporis Supputatione*, n. 315, 3. Cf. Vromant, *Normae Generales*, n. 138.

[96] Cf. *Epitome*, I, n. 728, and especially "A Quonam Momento Estne Professo a Votis Temporariis Integrum Religionem Deserere ad Normam Can. 637?,"—*Periodica*, XXII (1933), 34*-39*.—A similar doctrine is maintained by the following authors: Schäfer (*Compendium de Religiosis ad Normam Codicis Iuris Canonici* [2. ed., Münster i. W.: Aschendorffsche Verlagsbuchhandlung, 1931], n. 269); Cappello (*Summa*, II, n. 613; I, n. 181, 7, 3*); Chelodi-Bertagnolli (*Ius de Personis*, n. 89, b, 4, note 7; n. 273, note 3); Coronata (*Institutiones*, I, n. 592, 2); Blat (*Commentarium*, I, n. 96), and Toso (*Commentaria Minora*, I, 106). Many other authors are cited by Vermeersch (*Epitome*, I, n. 728).

There seems to be no canonical justification for such a stand. The only thing in its favor is that it exempts one from embarrassment.[97] It must be noted that the embarrassment that one might feel for having determined to leave the convent after one's temporary vows would not begin only on the day on which the perpetual profession would normally be made, but also a few days before this event. The question here under consideration is not one that is primarily concerned with the salvaging of possible disappointed human sensibilities, but one in which the fundamental concern is that of explaining and determining with the proper juridical limits the meaning and consequent application of a legal norm and standard. The word *renovandis* seems to indicate that n. 5 applies only to acts that are actually renewed. It would indeed be strange that the Code departed entirely from the rule that it so carefully enunciated concerning the determination of the starting point and left the duration of a time-interval, so to say, in the hands of private individuals. There seems to be no other justification for the special norm concerning the acts which are to be renewed save the fact that it contemplates the effective avoidance of possible difficulties, arising either from the intermission that could intervene between the renewal of such acts, or, conversely, from the overlapping of two supposedly successive acts. For, in the former instance the member of a religious community would be without vows during the intermission, while in the latter he would through the overlapping of his vows lack the desired interval of freedom before renewing his vows.[98]

[97] Vermeersch (*Epitome*, I, n. 728) writes: "Ceterum nonne expedit ut professus inconstans tunc abire permittatur cum eius confratres vota renovant, ne cum aliquo dedecore suo et aliorum tristitia manere debeat in domo postquam ipsius propositum omnibus innotuit?"

[98] Cf. canon 577, § 1. Vermeersch ("A Quonam Momento Estne Professo a Votis Temporariis Integrum Religionem Deserere ad Normam Can. 637?"—*Periodica*, XXII [1933], 34*-39*) produces other arguments with a view to proving his theory, but these may be rejected. He states, for instance, that the disposition of the law considers as a mere moment of time *(quasi unum fuerit momentum)* the day on which the profession should normally be made. No proof is adduced for this contention. Such

a limited duration of the day would be quite an innovation in law, and would certainly run counter to canon 32, § 1. Vermeersch maintains that his contention is confirmed by a reference, in the footnotes of the Code, to a decree of The Congregation of Religious under date of May 3, 1914. Well, the documents mentioned in the footnotes of the Code sometimes refer to laws which are directly opposed to the canons of the Code.

Another subtle argument is taken from the wording of n. 5 itself. Maroto and Michiels (Cf. *supra,* p. 226) use the text of n. 5 to show that the prescriptions therein contained are but an application of nn. 2 and 3 of canon 34, § 3. Vermeersch states that this cannot be the case on account of the words *"tempus finitur eodem recurrente die quo incipit"*. He maintains that there is no question here of the moment when the time-interval should normally begin to count according to canon 34, § 3, n. 3, but of the moment when the act to be renewed is posited. That contention is by no means justified by the text of the law. If the framers of the Code had in mind the moment when the act was posited, they could easily have found a clearer expression. The expression *"tempus finitur eodem recurrente die quo incipit"* must be understood in the light of Canon Law itself. The lapse of a specified period of time, in Canon Law, begins or expires according to definite rules, and not according to one's fancy. The present writer believes that the words under consideration could thus be interpreted: "The time expires with the recurrence of the day on which it began *juridically.*" And an interval of time is reckoned juridically, when there is a determinate starting point, according to the prescriptions of Canon 34, § 3, nn. 1 to 4.

CHAPTER IX

OTHER CONSIDERATIONS

ARTICLE I. AVAILABLE TIME

§ 1—Introductory Remarks

Available time is considered in canon 35: "*Tempus utile* means that time which for the exercise or prosecution of one's rights does not lapse if one was ignorant of one's rights, or was unable to act at the time; *tempus continuum* means that time which suffers no interruption." [1]

Available time, therefore, is the same as it was before the Code. No new provision has been added. The Code simply set down as a law what previously had been merely the common teaching of the doctors. The Code has not, however, determined precisely under which conditions time is to be computed, on the one hand, as available time, or, on the other, as time that runs with a continuous, non-intermittent juridical duration.

There are two well defined impediments which arrest the escape and stop the lapse of *tempus utile*: they are ignorance and the impossibility of using one's rights.

Ignorance, which simply means the absence of knowledge concerning one's rights, is easily determined in some instances, namely, when the Code states that the starting point of a period of time is reckoned from some official notification.[2]

Ignorance is naturally subjective, but the Code does not necessarily accept any and every kind of ignorance whatsoever as preventing the lapse of available time, but only such ignorance as a responsible man would be apt to have. If a person does not know what is obvious to all, he himself is held responsible for the loss of his rights.[3]

[1] Cicognani, *Canon Law,* p. 691.

[2] Canon 177, § 1: "Electus, si electio confirmatione indigeat, saltem intra octiduum a die acceptatae electionis confirmationem a competente Superiore petere per se vel per alium debet. . ."

[3] "Ignorantiam vero hic generatim accipimus, quae cadere possit in virum prudentem; quare, si quis quod omnibus perspectum est ignoret,

The ignorance here referred to encompasses a lack of knowl edge not only regarding the existence of one's rights, but also regarding the perpetrated violation of these rights.[4]

The word *agere* in canon 35 refers not only to the use of one's rights, or to their exercise but also to their prosecution. One exercises one's right for instance in an election by voting. The prosecution of one's rights has reference to court action. One prosecutes one's rights by defending them before a judge.

The meaning of the word *impediment* is very general and includes anything that can prevent one from using one's rights. The authors generally admit that such an impediment may legitimately arise even from one's own fault. Such is the case when one is suspended or excommunicated.[5] Indeed, an impediment may be either physical or legal. In the case of a suspension or of an excommunication it is merely legal. A physical impediment would be illness, lack of transportation facilities, etc.

There is a controversy as to the admissibility in Canon Law today of the *tempus utile ratione initii* which has already been mentioned.[6] Michiels thus presents the problem:

> Immerito ab aliquibus Auctoribus, ut CHELODI, p. 148, nota 3, et MATTHAEO A CORONATA, I, n. 56, p. 54, ad 4, et nota 1, adstruitur tempus quoddam *partialiter utile,* quod scilicet utile dicitur ratione initii tantum, in quantum v. g. ignoranti not currat, dum ratione cursus dicitur vere continuum. Sane non desunt canones, v. g. can. 430, § 2, 432, § 1, 1432 § 3, ubi statuitur jus vel obligationem non currere nisi a die acceptae notitiae, etc.; ex

sui iuris amissionem sibi imputare debet."—Toso, *Commentaria Minora,* I. 113. Cf. Van Hove, *De Temporis Supputatione,* n. 320; Cicognani, *Canon Law,* p. 692; Michiels, *Normae Generales,* II, 158.

[4] An example of this latter classification is found in the *Glossa* s.v. *sciverit*—c. 8, *de appellationibus,* II, 15, in VI°: "Et intelligo sciverit, non solum actum illum factum; sed per illum se gravatum. Quid enim si scivi te electum, et post decem dies scivi prius mihi de illo beneficio Papa providerat? Adhuc infra decem dies a die scientiae provocabo; quia modo scio me gravatum, et hoc sonat haec litera."

[5] Cf. canon 1654, § 1; D'Annibale, *Summula,* III. n. 35; Vermeersch-Creusen, *Epitome,* I, n. 151; Van Hove, *De Temporis Supputatione,* n. 320.

[6] Cf. *supra,* pp. 69-70.

> hoc tamen immerito deducitur tempus illud dicendum esse, ratione initii, utile. Tempus utile enim subaudit, facultatem vel obligationem illi, qui agendum esse nesciverit, ex benevolentia non currere a termino a quo, qui *objective et absolute* praedefinitus fuerit, sed a momento quo obligationem vel facultatem ipsi datam esse noverit; e contra, in canonibus ab illis Auctoribus adductis statuitur, quod *notitiae dies* constituit ipsum terminum a quo, cui objective et absolute ligatur obligationis vel facultatis initium.[7]

This controverted question does not seem to be of any practical importance. It is merely a question of the proper understanding of the expressions *tempus utile,* and *tempus utile ratione initii, continuum ratione cursus.* In canon 35 two impediments are listed which prevent the lapse of *tempus utile*: ignorance and the impossibility of using one's rights. These two are distinct and, theoretically at least, can exist separately. If, therefore, the Code conceded a certain interval of time in such a manner that it necessarily began to lapse in a continuous way from the very moment when one was informed of one's rights, that would be an example of *tempus utile ratione initii et continuum ratione cursus.*[8]

This seems to be precisely the case in canon 432 for instance. It is there stated that the Chapter is given eight days for the election of a Vicar Capitular *"ab accepta notitia vacationis"*.[9] In the first place it is certain that these eight days are not *tempus utile* in the full sense of the word, since it is also stated that if, for any reason whatsoever, the nomination has not been made within the prescribed time, it devolves on the Metropolitan. In practice, therefore, the election of the Vicar Capitular must be made within eight days of continuous time after notice has been received of the vacancy of the bishopric.

[7] *Normae Generales,* II, 160, note 2.

[8] Cf. Van Hove, *De Temporis Supputatione,* n. 318: "Nihil tamen obstat quin in iure canonico concedatur tempus utile ratione initii et ratione cursus, aut ratione initii tantum aut ratione cursus tantum. Quod si tempus utile conceditur ob ignorantiam, erit concessum utile ratione initii, ob impotentiam, potest esse utile ratione et initii et cursus."

[9] Canon 432, §1.

If one considered nothing but the bare wording of the canon it would seem that there could be no question here of available time, not even *ratione initii.* The accepted notification would simply be the starting point. However, the meaning that lies behind these words indicates that this is not the case. After all the logical starting point is not so much the knowledge of the vacancy, but the vacancy itself which is the determining factor. This real determining factor is mentioned in canon 34, § 3, n. 3: *"Octiduum a vacatione sedis episcopalis."*

If one accepted this notification as the starting point, one would be at a loss to define the nature of the starting points mentioned in canon 34. The starting points, as determined by law, cannot depend on the will of individuals for their efficacy. An implicitly determined starting point consists in a fact or event from which time is necessarily reckoned.[10] It is to be noted that nowhere in the Code can one find a definition or determination of the expression *"accepta notitia"*. The news of the bishop's death might not be confirmed, it might be doubtful, etc. Such might easily be the case if a bishop died in a city distant from his own episcopal city and in a region where communications are still in a rudimentary stage.

It seems therefore logical to conclude that the real starting point, of which there can be no doubt, is the bishop's death, and not the vague and, at times, all too uncertain notification of the vacancy. The expression used in the Code simply wishes to stress the point that no other impediment than that of improper knowledge is to prevent the eight days from lapsing.[11]

§ 2—The Reckoning of Available Time

Two distinct problems are here to be examined. The first considers available time in general and determines the mode of reckoning when no impediment occurs. The second examines

[10] Cf. *supra*, pp. 202-203.

[11] The existence of *tempus utile ratione initii, continuum ratione cursus,* is admitted also by Cicognani (*Canon Law,* p. 693), Wernz-Vidal (*Normae Generales,* n. 251), besides Coronata (*Institutiones,* I, n. 56 (p. 54 note 1) and Chelodi-Bertagnolli (*Ius de Personis,* n. 89. c. note 1).

those cases in which hindrances actually occur and impede the agent from acting. The reckoning to be made in the latter case is concerned with the computation of the unavailing days.

In the first place there seems to be no difficulty in the reckoning of available time when no impediment occurs. The days are then *de facto* continuous and follow the general rules set down in canon 34. In theory they might follow either the natural or the civil reckoning. In practice, however, the starting point is necessarily determined, since it would be impossible for the Code to grant a certain amount of available time without determining in some way the moment or act from which it is to be reckoned.

This seems to be the common teaching of the canonists. Oesterle is possibly the only author who mentions the possibility of reckoning available time from moment to moment.[12]

The reckoning of available time presents a problem of a totally different nature when impediments actually accur. One might indeed wonder how the unavailing days are to be computed. Van Hove states that the reckoning of available time is always according to the civil computation and that the *dies terminus a quo* is never reckoned as one of the days of the interval of available time. The reason for this is that the starting point in the reckoning of available time is always determined implicitly by the law. It is the moment when ignorance or other hindrances cease preventing one from making use of one's rights. It is furthermore to be noted that ignorance or other impediments do not customarily cease exactly at midnight.

Van Hove is also of the opinion that this form of the civil reckoning according to the prescriptions of canon 34, § 3, n. 3,

[12] "Iuxta Maroto, *l.c.* n. 260, 4, computatio temporis utilis non fit de momento in momentum, sed de die in diem, quia terminus a quo sive explicite sive implicite assignatus est (c. 34, § § 2 et 3). Puto: sic potest esse, sed non debet sic esse. Fieri posse mihi videtur, quod pro tempore utili terminus neque explicite neque implicite assignatus sit." *Praelectiones,* I, 21. Cf. Vermeersch-Creusen, *Epitome,* I, n. 151; Maroto, *Institutiones,* I, n. 260, 6°; Michiels, *Normae Generales,* II, 160-161; Vromant, *Normae Generales,* n. 139; Wernz-Vidal, *Normae Generales,* n. 251; Coronata, *Institutiones,* I, n. 56, 3; Van Hove, *De Temporis Supputatione,* n. 321.

is not only to be used in the determination of the beginning of availble time, but also in the computation of the unavailing days when available time has already begun to lapse.[13]

This statement, taken as a whole, seems to be correct only insofar as it prescribes the use of canon 34, § 3, n. 3, for the reckoning of a continuous period of available time. Thus in canon 1709, § 3, if the only impediment were ignorance, the ten days for recourse would actually begin to lapse from the moment of the notification and not from the moment of the rejection of the libellus. The ten days would then be reckoned continuously once the time began to lapse. Conversely one might suppose that two available days had already lapsed from the time of notification, and then another impediment arose. The agent, let us suppose, falls into a swoon at 8 A. M. July 16, which is the third available day, and does not recover before 6 P. M. of the same day. This is certainly a blameless impediment, and one which does not coincide with the beginning of the day. Common sense considers this day as of no avail from the juridical point of view. Correspondingly an extra day of available time should be added to the remaining seven days, which eight days together if they lapse as available time, will then fill out the total number of the ten days of available time.

One cannot come to this conclusion if one applies canon 34, § 3, nn. 2 and 3. That day, July 16, would not then count as an unavailing day, since the beginning of the impediment does not coincide with the beginning of the day. The reckoning of unavailing days would begin only on the following day. However, in the case here under consideration the following day, July 17, is fully available since there is no impediment.

The key to the solution lies in the fact that the use of canon 34, § 3, nn. 2 and 3, is intended *per se* for the computation of time-intervals which are necessarily continuous (*tempus de iure continuum*). At most one could argue that these prescriptions are also to be applied for the reckoning of intervals that are merely

[13] *De Temporis Supputatione,* n. 321. Cf. Maroto, *Institutiones,* I, n. 260, 4-6; Michiels, *Normae Generales,* II, 160-161.

de facto continuous. The use of canon 34, § 3, nn. 2 and 3, for intermittent periods would imply contradiction.

The wording of canon 34, § 3, nn. 2 and 3, seems to determine the beginning or the first day of time-intervals. This is misleading. The actual beginning of the duration of the various intervals reckoned according to these prescriptions is always determined by a physical fact or by the law itself. Thus, if a sentence is given at 10 A. M., July 10, the first juridical day of the ten within which an appeal may be made is July 11.[14] It is nevertheless obvious that an appeal may be made immediately after the sentence on July 10. Likewise, if one becomes a novice at 8 A. M., April 4, the first juridical day for the reckoning of the year required for the novitiate is April 5, but it would seem strange to say that the person did not become a real novice before April 5. The actual rôle of canon 34, § 3, nn. 2 and 3, consists in the determination of the end of the various time-intervals under consideration. Thus, in the case of appeal, canon 34, § 3, n. 3, rules that when a sentence is given at 10 A. M., August 12, for instance, the last moment of the time within which one may appeal is midnight April 22-April 23. It is therefore obvious that the prescriptions of canon 34, § 3, nn. 2 and 3, cannot be applied when an interval is *de facto* intermittent.

The problem therefore must be solved according to juridical principles other than those of canon 34, § 3, nn. 2 and 3. It is, however, equally certain that canon 34, § 2, is not to be applied, since the starting point is always at least implicitly determined, whether it be that of the original period of available time, or that of the starting point for the reckoning of unavailing days.

One must necessarily fall back on canon 32, § 1, for the reckoning of the day. Thus the ten days for recourse mentioned in canon 1709, § 3, for instance, are theoretically ten units of 24 hours, and nothing else. The question to consider is whether or not these ten days are indivisible units. If the days are divisible units, they may be divided into hours, minutes, etc. The ten days would then be the equivalent of 240 hours, and a person impeded two hours one day, six another, and so on, would count the hours

[14] Cf. canon 1881.

of impediment and add them to the original period granted by law. On the other hand, if the days are indivisible units, there is no question of counting the hours but merely of counting the days. A person is impeded, juridically, either for a full day or not at all. Thus, if a person were impeded from 8 A. M. to 10 P. M., a full day would be counted as being of no avail and a full extra day would be added to the ten originally given.

This is the common teaching of the authors. The Code indeed does not consider fractions of a day.[15]

The solution of the problem appears as soon as one knows what constitutes an impediment of one day's duration. A day, in theory, consists of 24 hours reckoned continuously from midnight to midnight. However, when available time is considered, then a day consists *de facto* in the amount of time which one may actually use, and varies with the nature of the institute involved. Thus a day for the purpose of appealing lasts only as long as the court or chancery office is open. It is obvious that one cannot appeal at 2 A. M. If these offices are open from 9 A. M. to 5 P. M., only eight hours are available. If one is impeded during these eight hours it is clear that a full canonical day is rendered unavailing and that an extra day is to be added to the total of the continuous days originally granted.

In other instances the day might be conceived as a little longer. Canon 155 rules that vacant ecclesiastical offices, apart from the cases in which a special law has set specific temporal limits for the making of the appointment, are to be filled with properly appointed incumbents within six months of available time to be reckoned from the moment when the vacancy becomes known. In this matter the ordinary could use practically any part of his waking hours for the proper provision in reference to these offices.

Canonists do not always make the proper distinction concern-

[15] Cf. *supra*, pp. 126-128; Maroto (*Institutiones*, I, n. 260, 6) writes: "Dies, in utili tempore computandi, videntur omnes et singuli debere esse completi, sive quia nunc dies manca non habetur umquam pro integra, nec incepta pro completa (supra n. 257, reg. 5, a), sive quia tempus utile continet favorem agentis qui est potius ampliandus."

ing the nature of the institute involved and simply state that the impediment need not last throughout the full physical day, but that it is sufficient that the hindrance last through a notable part of the day.[16]

"The amount of time," writes Kealy, "within the day necessary to constitute a notable part of it is not easily determined. Canonists are silent on the question, leaving it perhaps to the discretion of the judge in matters such as the one under present discussion [canon 1709, § 3]."[17] This interpretation, according to the same author, can easily lead to the assertion that, for certain persons, every day is impeded. That would be the case for those who are employed through the ordinary working day.[18]

Kealy then proposes a different opinion, on which however he does not seem to insist. He writes:

> A stricter opinion is that in view of the ease with which recourse may be made any solid opportunity for so doing during the day should be counted against the agent, and if he allows the day to pass without using his opportunity the day should not be regarded as impeded even though he may have been hindered during several hours of the day. One who holds this opinion might argue that a person occupied for six or eight hours in the day and unoccupied and unimpeded for the remainder of his waking hours has had sufficient opportunity to have recourse on that day and, hence, the

[16] Cf. Van Hove, *De Temporis Supputatione*, n. 321; Cappello, *Summa*, I, n. 182, 1; Coronata, *Institutiones*, I, n. 56, 3; Wernz-Vidal, *Normae Generales*, n. 251.—It would seem pertinent to add, however, that the notable part of the day here referred to is not the theoretical day of 24 hours, but only that part of it within which the business involved may be transacted. If the words "a notable part of the day" referred to the 24 hour day, the logical conclusion to draw would be that available time would never lapse since one's sleep, meals, etc. necessarily impede one through a notable part of every day of the year.

[17] *The Introductory Libellus in Church Court Procedure*, The Catholic University of America, Canon Law Studies, n.108, (Washington: The Catholic University of America, 1937), p. 70.

[18] *Loc. cit.*

day should be counted as one of the ten allowed by the law.[19]

It is however to be remarked that the fact that one is working is not necessarily considered an impediment in the full sense of the law. Many recourses or appeals may easily be made by one's procurator or otherwise. A person alleging his work as a hindrance would most likely not be sincere. Against the opinion proposed by Kealy there seems to be the text of the law itself, according to which the person has ten full availing days to act. One could therefore logically conclude not only that the impediment must be real and not imaginary, but also that, when it does exist, it is sufficient that one be hindered merely through a notable part of the day.

A more precise statement concerning what constitutes a notable part of the day cannot here be given. It depends, first of all, on the nature of the institute involved, but simultaneously also on a limitless number of accompanying circumstances which a prudent judge or superior must consider and weigh in determining whether or not this or that day is to be considered as being juridically of no avail. In fine, the judgment in such cases rests upon moral considerations.

Van Hove maintains that available time is not reckoned morally, but physically, like the other time-intervals. He states that the only moral consideration in the reckoning of *tempus utile* lies in the determination of the value or nature of the impediments.[20]

This does not give a full explanation of the problem. Two things are involved in the reckoning of *tempus utile,* namely, the moral computation and the moral judgment of the character of the impediment. Thus, if one is unconscious from 7 A. M. to 3 P. M., and absolutely normal the rest of the day, there is no question of using a moral estimation for the determination of the impediment. That is already a physical certitude of which there can be no doubt. On the other hand, if one says that a person certainly

[19] *Loc. cit.*

[20] *De Temporis Supputatione*, n. 321.

impeded for six hours during a day is entitled to a full extra day to be added to the number originally granted by the law, it becomes obvious that a moral reckoning is used. *Tempus utile* necessarily contains the notion of a variable adaptability in view of diversified circumstances in divergent cases and that is precisely the situation from which the essence of the moral reckoning must be drawn.

§ 3—The Occurrence of Available Time

Time is of its nature continuous. It lapses whether or not the person to whom it is granted is able to use it. There must always be some special reason before one may rightly regard a definitely fixed time-period as one of available time. A few explicit legal prescriptions to this effect are found in the Code. In other instances available time is said to be conceded implicitly when one is given a certain amount of time to accomplish a certain act which is looked upon as favorable to the agent and which, at the same time, is odious to no one.[21]

Tempus utile in the full sense of the expression as explained in canon 35 is explicitly mentioned in canon 155, according to which the canonical provision to be made for an office, if no definite time has been prescribed, is not to be deferred for more than six months of available time, and these are to be reckoned from the time that the authority which is to make the provision has received notification regarding the vacancy. The available time granted for effecting this canonical provision in respect of the vacant ecclesiastical office is the same also in relation to the appointment of incumbents in vacant ecclesiastical benefices. This must be concluded in virtue of canons 146 and 1413, § 2, according to which the canons that treat of ecclesiastical offices are to

[21] Cf. Cance, *Le Code de Droit Canonique,* n. 75, c; Cappello, *Summa,* I, n. 182, 2; Chelodi-Bertagnolli, *Ius de Personis,* n. 89, c; Cicognani, *Canon Law,* p. 692; Cocchi, *Commentarium,* I, n. 98; Coronata, *Institutiones,* I, n. 56, 2; Maroto, *Institutiones,* I, n. 260, 2; Michiels, *Normae Generales,* II, 160; Toso, *Commentaria Minora,* I, 114; Van Hove, *De Temporis Supputatione,* n. 319; Vermeersch-Creusen, *Epitome,* I, n. 151; Vromant, *Normae Generales,* n. 139; Wernz-Vidal, *Normae Generales,* n. 251.

be applied also with reference to ecclesiastical benefices. Therefore the six months of time which canon 1432, § 2, grants to the ordinary in the conferment of a vacant ecclesiastical benefice are to be considered as six months of *tempus utile*.

Another instance of express mention is found in canon 161, in which it is stipulated that when an election is conducted by a collegiate body, this must be done within three months of available time to be computed from the time that notification has been received concerning the vacancy of the office to which one is to be elected. One can imagine the case of a notification received June 30. If there is no impediment the three months lapse as continuous time and are to be reckoned according to the calendar. Furthermore, the first day does not count.[22] Consequently the time-interval here under consideration would expire at midnight September 30-October 1. On the other hand, if an impediment arises once the time-interval has begun to lapse, then the three months of available time do not run as continuous time but as intermittent time. It would seem in that case that the three months should not be reckoned according to the calendar, but should be given an aggregate total of 90 days. There is no direct canonical evidence to prove this contention, since canon 34, § 3, considers *per se* only those time-intervals that lapse without interruption. Nor can canon 34, § 2, be applied directly since the provisions of § 2 are concerned only with those instances of time-reckoning in which there is no determinate starting point. It has already been shown that the cessation of an impediment constitutes a determinate starting point with a view to computing the remainder of the available time. But, in view of canon 20 it would seem altogether relevant to compute any interval running as intermittent time not according to the calendar, but according to the juridical reckoning in which the month and the year always have 30 and 365 days respectively.

Clear instances of *tempus utile* without restriction as to the nature of the impediments are also found in canons 175, 1457, and 1709, § 3.[23]

[22] Cf. canon 34, § 3, nn. 1 and 3.

[23] Cf. Maroto, *Institutiones,* I, n. 260; Van Hove, *De Temporis Sup-*

It is to be noted that express concessions of available time may be made without the explicit use of the words *tempus utile*. In many instances the context clearly shows that a certain period, which in itself seems to be continuous time, is really available time, because the Code says elsewhere that the time does not expire if an impediment prevents the agent from using his rights. An example of this is found in canon 177, § 1.[24] Another instance of this is had in canon 181, § 1.

At times the Code uses the words *tempus utile,* but not in the meaning of canon 35. Thus from canon 1902, n. 2, one learns that the *res iudicata* is had if no appeal is made against the sentence of the first tribunal within the prescribed "*tempus utile*". This simply means that the *res iudicata* exists if there is no appeal within ten days, but these days do not seem to be available time. The expression *tempus utile* in this instance simply refers in general to the time that one could use, and that is ten days of *tempus continuum.*[25] Canon 1884, § 2, presents a similar situation.

There is some doubt as to the nature of the *tempus utile* mentioned in canon 1702: "Omnis criminalis actio perimitur . . . lapsu temporis utilis ad actionem criminalem proponendam". The reason for doubt comes from canon 1705, § 1, where it is ruled that the starting point for the prescription of a criminal case is the perpetration itself of the delict, as opposed to the starting point for contentious causes, which is the moment when the matter could be brought to the attention of the tribunals.[26] The present writer is of the opinion that the *tempus utile* mentioned in canon 1702 is not real available time in the sense of

putatione, n. 319; Vermeersch-Creusen, *Epitome,* I, n. 151; Vromant, *Normae Generales,* n. 139; Wernz-Vidal, *Normae Generales,* n. 251.—The year mentioned in can. 2340, § 1, is also to be regarded as a year of available time. Cf. *Apollinaris,* VIII (1935), 410.

[24] "Electus, si electio confirmatione indigeat, saltem intra octiduum a die acceptatae electionis confirmationem a competente Superiore petere per se vel per alium debet; secus omni iure privatur, nisi probaverit se a petenda confirmatione iusto impedimento fuisse detentum."

[25] Cf. Roberti, *De Processibus,* II, n. 478, 1 and 510, b.

[26] Cf. Van Hove, *De Temporis Supputatione,* n. 319.

canon 35. The *tempus utile* of canon 1702 simply refers to the time-interval one can *use* according to the prescriptions of canon 1705, § 1.

The Code makes no reference to the time-period during which recourse may or must be made to the Holy See as granted by canon 647, § 2 n. 4, to religious professed with temporary vows if they receive a decree of dismissal from their superiors. But the Holy See in 1923 decreed that this recourse may and, if it is to have juridical effect, must be made within ten days, and that this period of time is to be reckoned as one of available time.[27]

The case of canon 2146 is similar. The Holy See decreed that the recourse here mentioned had to be made within ten days of *tempus utile*.[28]

Van Hove also designates as available time the four years that the Code grants to minors for petitioning the *restitutio in integrum*.[29]

A word could also be said here about a secondary or mitigated form of available time. *Tempus utile* as defined in canon 35 considers any impediment in general as capable of arresting the lapse of available time. In certain cases, however, the Code restricts the number of these impediments so that only one determinate hindrance may be canonically recognized as stopping the flow of available time in its otherwise continuous course. It has already been shown that this is the case when the Code uses the words *a recepta notitia* or some other similar expression provided, however, that the allotted time-period is not for any specifically

[27] S. C. de Relig., *Decretum*, 20 iulii, 1923—*AAS*, XV (1923), 457: "The available time (*tempus utile*) for the interposition of the recourse as regards the suspensive effects mentioned in can. 647, § 2, is ten days from the notice to the dismissed religious, according to the norm established in similar cases, as in cc. 1465, § 1, and 2153, § 1."—Bouscaren, *CLD*, I, 328.

[28] S.C.Conc. *Resolutio*, 14 ian. 1924—*AAS*, XVI (1924), 162; Bouscaren, *CLD*, I, 837.

[29] *De Temporis Supputatione*, n. 319: "Docent scriptores quadriennium concessum ad petendam restitutionem in integrum esse computandum utile, quod iam sub iure Iustiniano erat admissum."—Cf. Vermeersch-Creusen, *Epitome*, III (3. ed.) n. 120; C. (II, 52 [53]) 7.

expressed reasons designated as a *tempus utile*. In some instances, indeed, the Code grants available time without mentioning the words *tempus utile*. Such is the case when a time-interval is to be reckoned from some official notification, and is furthermore said not to lapse if there be a legitimately acceptable impediment.[30]

A single impediment other than ignorance is mentioned in canon 1635, according to which the day whereon a juridical act is to be posited does not count if the tribunal does not function on that day.[31] Thus, if the tenth day of the period during which an appeal must be made happens to be a court holiday, this circumstance prevents the tenth day from lapsing and an extra day is added to the total originally granted by the law. A similar case occurs in canon 1884, § 2, which rules that if one is unable to obtain, within the *tempus utile*, a copy of the impugned sentence from the tribunal of the first instance, the time does not lapse in the meantime, and the impediment is to be called to the attention of the higher tribunal. The words *tempus utile* are here textually inserted in the canon, but the interval referred to is nonetheless only partially equivalent to that of the *tempus utile* defined in canon 35. The only impediment that prevents the prescribed time from lapsing is the impossibility of obtaining from the tribunal of first instance a copy of the sentence there given.[32]

Article II. The Moral Reckoning of Time

A question very closely associated with that of available time

[30] Cf. *supra*, pp. 231-233. This is the question of the *tempus utile ratione initii, continuum ratione cursus*, which is not admitted by many authors.

[31] "Si dies, pro actu iudiciali indicta, sit feriata nec in decreto iudicis dicatur expresse tribunal vacaturum nihilominus causis cognoscendis, terminus intelligitur prorogatus ad primam sequentem diem non feriatam."

[32] Canon 1884, § 2 reads: "Quod si pars exemplar impugnatae sententiae intra utile tempus a tribunali *a quo* obtinere nequeat, interim termini non decurrunt et impedimentum significandum est iudici appellationis, qui iudicem *a quo* praecepto obstringat officio suo quamprimum satisfaciendi." Cf. Cicognani, *Canon Law*, pp. 693-694, where canon 432, § 3, is examined.

is that of the moral reckoning. Under this caption the writer wishes to consider certain principles which were held in great esteem by the canonists before the promulgation of the Code, namely, *parum pro nihilo reputatur* and *dies incepta habetur pro completa.*

§ 1—*Parum pro Nihilo Reputatur*

This old axiom is rejected by many authors in a very general way.[33] Their main argument lies in these words which seem to be common property among the pre-Code writers: "Nam in his quae natura, aut jure definita sunt, non est morali aestimationi locus."[34]

The so called moral reckoning consists precisely in the use of the principle *parum pro nihilo reputatur,* or in the use of a more or less flexible standard in the computation of time. The expression *moral reckoning* is often used in Canon Law, but not always in the sense here considered.[35] Cicognani, for instance, considers as moral the reckoning of the age required for the reception of the Sacred Orders under the law of the Council of Trent.[36]

The expression *moral reckoning* as the present writer understands it is a misnomer. It is not a reckoning at all in the strict sense of this word, but rather a norm which guides one in the concrete application or use of time-reckoning in general. Canon 788, for instance, states that the age prescribed for the administration of Confirmation is about seven years.[37]

In this case the seven years are not to be reckoned in any way whatsoever according to one's option. One is bound to say that

[33] Cf. Cicognani, *Canon Law,* p. 682; Lacau, *De Tempore,* pp. 33-34; Maroto, *Institutiones,* I, n. 257, 1 and 5; Michiels, *Normae Generales,* II, 157.

[34] D'Annibale, *Summula,* I, n. 39. Cf. Maroto, *op. cit.* n. 257, 1.

[35] Cf. *supra,* p. 89.

[36] *Canon Law,* p. 683, where this reckoning is contraposed to the strict and the physical computation.

[37] "Licet sacramenti confirmationis administratio convenienter in Ecclesia Latina differatur ad septimum circiter aetatis annum, nihilominus etiam antea conferri potest, si infans in mortis periculo sit constitutus, vel ministro id expedire ob iustas et graves causas videatur."

a child is seven years old when the seven years, reckoned according to canon 34, § 3, n. 3, have lapsed. On the other hand, there is practically no obligation of waiting until the child has completed these seven years according to this reckoning before the Sacrament may be administered. Any justifiable cause may warrant action a few days, a few weeks, or even a few months or years before that age is actually reached. Such is the necessary conclusion to draw from canon 788, which indicates a certain age, more or less as an example, without imposing an obligation to adhere to it in an absolute, physical way.

It is also obvious that the principle *parum pro nihilo reputatur* is admissible in the reckoning of available time. It was shown above that a certain impediment lasting only for a very short time is not considered. There is no doubt that the moral reckoning is to be used in many instances throughout the Code. Such is the case in canon 1620, where the Code imposes the obligation on judges and tribunals of finishing causes in the shortest time possible. In the court of first instance cases should not be prolonged over two years; in the court of second instance one year is set down as the limit. However, many circumstances may prolong cases beyond these limits. Even if there were no just cause for such a prolongation, no rights would be lost.[38]

Many authors hold that the age of puberty in penal affairs is fourteen years for women as well as for men.[39] According to this opinion, therefore, one must necessarily have recourse to the moral reckoning to compute the twelve years of age mentioned in canon 88, § 2, with reference to the age of puberty in women.

The writer's contention is that the canons which deal with the reckoning of time merely consider the abstract, theoretical duration of the time-periods involved, and do not apply to the moral obligation that one has of adhering strictly to them. This obligation depends on the nature of the various canons throughout

[38] Cf. Roberti, *De Processibus*, I, n. 180. Cf. also canons 867, § 4 (nisi aliud rationabilis causa suadeat); 399, § 1.

[39] Cf. canons 2230; 2218, § 1. This opinion is maintained by Reiffenstuel (lib. V, tit. XXIII, n. 5); Lega (*De Delictis et Poenis*, n. 37). Cf. Vermeersch-Creusen, *Epitome*, III (3. ed.), n. 424; canon 1648, § 3.

the Code where time-intervals are found mentioned. If canons 31-35 dealt with this, the same moral obligation would theoretically be imposed in all the instances throughout the Code where time-reckoning is involved. And the violation of a few minutes in the reckoning of a hundred years would be just as serious as the violation of a similar amount in the reckoning of a single day.

If the reckoning and the use of a time-interval were one and the same thing, then one could not explain canon 34, § 3, n. 3. According to this canon the day on which an act is performed does not count in the reckoning. Thus a postulant might become a novice at 9 A. M. July 15. However, the first day for the reckoning of the year of the novitiate is July 16, yet no one will deny that the postulant in question becomes a full-fledged novice at 9 A. M. July 15, with all the rights and obligations annexed to that state of life.

A certain number of authors favoring the physical and strict reckoning of time give a few examples as illustrative proof of their contention. Maroto, for instance, after stating as a general rule that time must be reckoned physically, goes on to mention a few cases in which the rule applies. Such is the case for the reckoning of time in the observance of the eucharistic fast, in the observance of abstinence days, in the reckoning of the age of twenty-one with reference to fasting, in the reckoning of age relative to the impediment of age in the reception of orders or in the contracting of valid marriages.[40]

The present writer does not for an instant doubt that the time must be reckoned physically in the above mentioned cases. But it is a different thing to say that the application of the physical reckoning in these cases is imposed by canons 31-35 and not by the respective nature of the various institutes under consideration.

A few authors have with reason taken a different view on the matter and state that time is to be reckoned physically, not in every case, but when it pertains to the form of an act. Toso writes: "Quare tempori, *quando datum est pro forma, videlicet*

[40] *Institutiones,* I, n. 257, 1. Cf. Michiels, *Normae Generales,* II, 157; Lacau, *De Tempore,* pp. 33-34; Cicognani, *Canon Law,* p. 682.

quando petitum est ad aliquem actum, nec minimum quid addi aut detrahi potest (S. Rota, in *Recentior.* decis. 300, n. 29 et 31 part. 10)."[41] Time pertains to the form of an act at least in those instances where the validity of an act is involved.

What has been advanced thus far in this article concerning the moral reckoning of time is confirmed in some measure by a private decision of the Code Commission of Interpretation with reference to canon 858, § 2. It is stated in this canon that the faithful who have been ailing for a month and who have no certain hope of recuperating within a short time may, after consulting a prudent confessor, receive Holy Communion once or twice a week, even after taking medicine or food in the form of a beverage.[42]

The majority perhaps of the authors maintain that the month here mentioned is not by any means to be reckoned morally, but physically, so that a person must actually be sick for a full month according to canon 34, § 3, n. 3, before this privilege may be invoked.[43]

Before the Code, however, the more common opinion held that this month was to be reckoned morally, so that a person was considered capable of using this privilege after 26 or 27 days of illness. This opinion is now held in substance at least by Cappello.[44]

[41] *Commentaria Minora,* I, 103. Cf. Chelodi-Bertagnolli, *Ius de Personis,* 89, b, 4; Cappello, *Summa,* I, n. 181, 8, where he mentions the reckoning of the age for matrimony as an example in which the computation of time pertains to the form of the act.

[42] Canon 858, § 2: "Infirmi tamen qui iam a mense decumbunt sine certa spe ut cito convalescant, de prudenti confessarii consilio sanctissimam Eucharistiam sumere possunt semel aut bis in hebdomada, etsi aliquam medicinam vel aliquid per modum potus antea sumpserint."

[43] Cf. Jorio, *La Comunione agl'Infermi,* nn. 64-66; Kinane, "Some Queries in Regard to the Modification of the Eucharistic Fast for the Sick,"—*IER,* XXXV (1930), 519-521; Van Hove, *De Temporis Supputatione,* n. 280, note 2; Beste, *Introductio,* p. 100.

[44] Cf. *De Sacramentis,* I (1921), n. 506, where explicit mention is made of 26 or 27 days. In the third edition of the same work (*De Sacramentis,* I [1938], n. 506, 2) the moral reckoning is vigorously and rightly

A private decision of the Code Commission of Interpretation under date of Nov. 24, 1927, declared that the month mentioned in canon 858, § 2, was not to be reckoned mathematically, but rather morally.[45] This answer is in perfect agreement with the prescriptions of canon 6, n. 2, according to which the "canons of the Code that restate former laws exactly as they were before, must be interpreted according to the approved and accepted interpretation of commentators on the old law."[46]

One thing is certain from what has been seen. In theory, the month here referred to is to be reckoned according to canon 34, § 3, n. 3, but it can in no way be deduced from this canon that one must actually be ill for at least 28, 29, 30 or 31 days as the case may be, before this privilege may be used. Canon 34 does not even consider the problem from that angle.

A problem closely associated with the one here under consideration is found in canon 1115, § 2. According to the prescriptions of this canon "the children who are born at least six months after the date of marriage, or within ten months from the dissolution of conjugal life, are in law presumed legitimate".[47] Theoretically at least, the two time-intervals here mentioned should be reckoned according to the calendar in view of canon 34, § 3, n. 1.[48] However, a few canonists propose a different computation in practice. Van Hove states that extraplacental life becomes possible after 180 days of gestation, that is, after 6

upheld by Cappello, but no explicit mention is made of 26 or 27 days. Cf. also Twomey, "The Eucharistic Fast,"—*ER*, CII (1940) 418. This author also mentions the following writers as maintaining a similar opinion: Tummolo-Iorio, *Compendium Theologiae Moralis*, n. 338; Ubach, *Theologia Moralis*, n. 1784.

[45] Cf. Bouscaren, *CLD*, II, 88. Cf. also *Periodica, XXIII* (1934), 234* (Cappello); Vermeersch, "Recta computatio mensis, quo elapso, licet semel vel bis in hebdomada, decumbenti infirmo permittere ut s. dape reficiatur postquam aliquid per modum potus vel medicinae sumpserit (can. 858, § 2).,"—*Periodica*, XXIII (1934), 61*-63*.

[46] Woywod, *The New Canon Law* (new ed., New Yoık: Wagner, 1918), p. 2. Cf. Cappello, *De Sacramentis*, I (3. ed.), n. 506, 2.

[47] Woywod, *The New Canon Law*, p. 227.

[48] Cf. Van Hove, *De Temporis Supputatione*, n. 313.

months of 30 days. Therefore, he adds, the six months mentioned in canon 1115, § 2, must be given the juridical duration of 30 days, and must not be reckoned according to the calendar because this form of reckoning would at times prove to be unjust.[49]

The present writer does in no way adhere to such a doctrine. The 180 days mentioned by Van Hove were adopted in the Roman Law on the authority of Hipprocrates (460-357 B. C.), but the opinions of the "Father of Medicine" are not necessarily to be held today. Children have been known to be born before 180 days of gestation. A case has been put on record in which a child was born after 158 days of gestation.[50] Furthermore it is often stated today that the period of possible extraplacental life is from the twenty-seventh week onward.[51]

Another remark might be made. The presumption mentioned in canon 1115, §2, is not a *praesumptio iuris et de iure,* but a *praesumptio iuris tantum.* In other words, the presumed legitimacy does not stand in the face of opposite facts. If a child is born 185 days after the date of marriage and is nevertheless considered fully developed by competent medical men, it is obvious that this child cannot be regarded as legitimate in the light of canon 1115, § 2. On the other hand, if a child is born 182 days after the date of marriage and the question of legitimacy is brought up many years later at a time when it is no

[49] *De Temporis Supputatione,* n. 313. Months of 30 days are also propounded by Vromant (*Normae Generales,* n. 137), De Becker (*De Matrimonio Praelectiones Canonicae* [ed. nova, Lovanii, 1931], p. 206), Ayrinhac (*Marriage Legislation in the New Code of Canon Law*—Revised and Enlarged by P. J. Lydon, D.D. [New York: Benziger, 1932], n. 279, 2). Van Hove (*De Temporis Supputatione,* n. 313) cites Vermeersch-Creusen (*Epitome,* II [4.ed.], n. 420) as maintaining a similar opinion. In the fifth edition of the *Epitome* Vermeersch-Creusen state that strange consequences might occur if the six months now under consideration are reckoned according to the calendar. These authors nevertheless state that the six months mentioned in canon 1115, § 2, must be reckoned according to the calendar. Cf. *Epitome,* II (5. ed), n. 420.

[50] Cf. Medicus, *Medical Essays* (3. ed., 1931), p. 314.

[51] Cf. *op. cit.* p. 316.

longer possible to have any definite knowledge concerning the state of the child's development at the time of birth, then the child is presumed legitimate if it was born six months after the date of marriage. And these months are to be reckoned according to the calendar in view of canon 34, § 3, n. 1. As a matter of fact canon 1115, § 2, is very lenient. In every-day life a child born six months after the marriage of his parents is simply looked upon as having been conceived outside of lawful wedlock.

De Becker maintains that the six months here under consideration must be reckoned as they were before the Code.[52] There is no canonical justification for such a stand. If the six months were reckoned according to the juridical computation so as to mean a period of not more than 180 days, that state of affairs has been radically changed by the Code which prescribes the calendar reckoning.

§ 2—*Dies Incepta Habetur pro Completa*

This principle no longer exists according to the meaning it had in the old law. Before the Code the last day of any enumeration in favorable matters was considered as a full day even though it was but begun. Thus a year from January 1 ended juridically with the beginning of December 31.[53]

Such reckonings are now excluded by the Code which prescribes well determined norms in this field. In exceptional cases, however, this principle may be said to survive, but in a slightly different sense. It has already been shown that, according to some interpretations of canon 34, § 3, n. 5, a day begun is held as complete.[54]

A day begun is held as complete also in the reckoning of available time. Thus one need not be impeded through 24 solid hours before an extra day may be added to the original number granted by the law.[55] Here, it is true, impediments of a minor

[52] *De Matrimonio Praelectiones Canonicae,* p. 206.

[53] Cf. *supra,* pp. 84-86.

[54] Cf. *supra,* pp. 226-227.

[55] Cf. *supra,* pp. 236-237.

duration are not considered. But when an impediment lasts long enough to be counted, it is immediately reckoned as implying the loss of a full day. Likewise, in the matter of vacations it has been shown that the day is not a divisible entity, and that if a pastor has been absent from his parish for a notable part of the day, a full day is to be subtracted from the amount of time that the law grants him for a vacation.[56] In that sense, therefore, it may still be held that a day though only begun is nevertheless held as a day already completed.

§ 3—The Completion of Time

Another point closely connected with the one just examined is that of the completion of time. In the old law there were many instances when a year or a month were held as completed the moment they began.[57]

This form of reckoning should possibly be examined from a different angle in the light of the Code, but it is safe to say that no fundamental change has been made in this matter. Thus it is obvious that when the Code mentions one, two or three years, these years are reckoned as full years according to canon 32, § 2. However, there might be a few expressions in the Code where a certain number is used, but in such a way as to mean the number just before it. Thus a person is not by any means six years old when he begins his sixth year. He is merely five years old. Such is the case of canon 1254, § 2: "Lege ieiunii adstringuntur omnes ab expleto vicesimo primo aetatis anno ad inceptum sexagesimum."[58]

The problem, in reality, is not whether the last year or month of any enumeration prescribed by the Code is to be reckoned as a full year or not, but rather to determine the exact number of years or months prescribed by the Code. In such instances, when a doubt arises, one is to refer to the old law. Thus in can. 766, n. 1, the question is whether or not a godfather must be thirteen or fourteen years of age. The expression *"attigerit"*

[56] Cf. *supra*, pp. 126-128.

[57] Cf. *supra*, pp. 80-84.

[58] Cf. also canons 766, n. 1 and 796, n. 3.

which is there employed, has always been accepted as meaning that the year explicitly mentioned was to be reckoned as completed the moment it began.[59]

Every time the Code uses a cardinal number in mentioning a certain number of months or years there is no doubt that the actual number indicated must consist of full months or years, unless there be an explicit provision to the contrary, or unless the time-interval under consideration be reckoned morally. Instances of full years are found in canons 331, § 1, n. 2; 367, § 1; 524, etc. On the other hand, when an ordinal number is used in designating a certain month or year, one must not necessarily conclude that the month or year actually mentioned is to be a full month or year. In such instances the complete text of the law is to be examined, and should be interpreted according to the common and approved teaching of the canonists.[60]

Such expressions as "*ad sexagesimum annum*", "*ante decimum octavum annum*" normally indicate that the last year mentioned need not be a full year. Conversely, the proposition "*post*" demands that the last month or year mentioned be a full month or year.

When an ordinal number is used, but with the expression "*expleverunt*" or any other similar or equivalent term, there is no doubt that the last year of the enumeration must be completed. Such an instance is found in canon 12, where it is stated that those who have not yet completed their seventh year are not bound by ecclesiastical laws, unless the contrary is explicitly mentioned. This really means that one is not bound by ecclesiastical laws before the beginning of one's eighth year. Seven full years must be completed. Similar instances are found in canons 88; 555, § 1 n. 1; 975; 1067, § 1. *A fortiori* full years are to be reckoned when cardinal numbers are used and the Code insists furthermore that these be full years, as in canons 27; 28; 504; 539, § 1.

The question, from what has been seen, is slightly different

[59] Cf. *supra*, p. 82; Maroto, *Institutiones*, I, n. 257, 5, note 2.
[60] Cf. *supra*, pp. 80-84.

from what it was in the old law. Before the Code in many instances even a cardinal number was to be so interpreted that the last year was to be considered as full the moment it began. This held especially for the reckoning of the 25 years required to hold civil honors. This form of time-reckoning no longer exists in Canon Law.[61]

[61] Cf. *supra*, pp. 81-82.

CONCLUSIONS

1. The wording of canon 31 has a wider latitude than what was most likely intended by the framers of the Code.

2. The word *calendar* must be taken in a broad sense so as to include, besides the Gregorian calendar, the liturgical, the scholastic, and various forms of local calendars.

3. A day in law is always a continuous period of 24 hours, whether the starting point be midnight (civil reckoning) or any other moment (natural reckoning). Fractions of a day are considered either as full days or as nothing at all.

4. Standard Time is always a legal time in the United States.

5. The liberal interpretation of canon 33, § 1, is undoubtedly held by the greater number of canonists at the present time. One may, without the least doubt, adopt this interpretation in practice.

6. The mental determination of following one of the time-reckonings mentioned in canon 33, § 1, has no canonical effect. The actual use of one or the other of the computations is alone considered.

7. In practice one may use the latitude granted by canon 33, § 1, for the observance of all precepts of a private nature.

8. The examples given by the Code do not permit a clear-cut conception of the determination of the starting point in reference to the use of the civil or the natural reckoning. There always seems to be a determinate starting point for the computation of the time through which a suspension as a vindictive penalty must be observed.

9. The natural reckoning seems to serve no useful purpose. This form of computation is not found in the German Code, and Canon Law would gain if it were abolished.

10. The words *continuous* and *intermittent* as applied to time-intervals in canon 34 § 2, refer only to periods *de facto* continuous or intermittent.

11. Canon 34, § 3, n. 5, is in all probability to be applied only to acts which are actually renewed. The prescription of n. 5

of canon 34, § 3, is an application of nn. 2-3 of the same canon.

12. The prescriptions of canon 34, § 3, are to be applied only to time-intervals which are at least *de facto* continuous.

13. There is no innovation in the Code concerning the reckoning of available time in general.

14. Canons 31-35 are concerned with the *reckoning,* and not with the *use* of time. In other words, the nature of the institute involved determines whether or not a time-interval is to be observed physically or morally.

15. The principle *dies incepta habetur pro completa* may no longer be invoked according to the meaning it had in the pre-Code legislation.

16. Years and months can no longer be held as completed the moment they begin unless the text of the law clearly indicates that a full year or month is not required.

APPENDIX I

Practical Tables for the Reckoning of Mean Solar Time and True Solar Time for any Given Community

This dissertation would certainly not be complete without practical information on how to find the various times one may licitly use.

In the course of this work mention has been made of usual time, of local time (true and mean) and of legal time (regional and extraordinary).

Usual time and legal time offer no difficulty. Mean solar time and true solar time, on the contrary, require a little computation. The following tables have therefore been constructed to enable those who wish to do so to find the time most advantageous to them for the carrying out of their various obligations. The tables should enable one to reckon the earliest and the latest "midnight". When midnight puts an end to an obligation, as with the law of abstinence on Friday evening, the earliest possible reckoning of midnight may be sought for, and vice versa, when midnight marks the beginning of a new obligation, as with the eucharistic fast, the latest possible reckoning of midnight may prove more acceptable.

TABLE I

Mean Time

Mean time is relatively easy to find. The first step is to determine the longitude of one's locality. If that city or town is situated east of the Standard Time meridians, namely the 75th, 90th, 105th and 120th, its mean midnight will be ahead of, or faster than, the midnight indicated on the Standard Time clocks and watches according to the rate of four minutes for each degree of longitude. Vice versa, if that community is situated west of the standard meridians, its mean midnight will be short of, or slower than, the midnight indicated on the Standard Time clocks and watches according to the rate of four minutes for each degree of longitude. Thus, Boston is on the 71st meridian, or four degrees east of the 75th. Multiply 4 by 4 minutes and you have the total of 16 minutes. It is therefore mean midnight in Boston at 11:44 P. M. according to a correct watch running on EST. On the contrary, Washington is on the 77th meridian, that is two degrees west of the 75th. Standard Time clocks in Washington are therefore eight minutes ahead of the mean sun time which registers midnight at 12:08 A. M. EST.

Since it is impossible to give the mean time of every town or city in the United States, the names of the episcopal cities will be recounted here. The following figures have been compiled from various maps and atlases. The asterisk (*) is used to denote a plus quantity and indicates that the Standard Time clock is fast by the number of minutes shown. The dash (-) denotes a minus quantity and indicates that the Standard Time clock is slow by the number of minutes shown.

Province of Baltimore-Washington

Baltimore, Md.	EST	* 6.5	minutes
Washington, D. C.	EST	* 8.0	"
Atlanta, Ga.	CST	–22.5	"
Charleston, S. C.	EST	*19.5	"
Raleigh, N. C.	EST	*14.0	"
Richmond, Va.	EST	* 9.5	"
St. Augustine, Fla.	EST	*22.5	"
Savannah, Ga.	EST	*24.5	"
Wheeling, W. Va.	EST	*23.0	"
Wilmington, Del.	EST	* 2.5	"
Belmont Abbey, N. C.	EST	*24.0	"

Province of Boston

Boston, Mass.	EST	–16.0	minutes
Burlington, Vt.	EST	– 7.0	"
Fall River, Mass.	EST	–15.5	"
Hartford, Conn.	EST	– 9.5	"
Manchester, N. H.	EST	–14.0	"
Portland, Me.	EST	–19.0	"
Providence, R. I.	EST	–14.5	"

Springfield, Mass.	EST	– 9.5 "
Province of Chicago		
Chicago, Ill.	CST	– 9.5 minutes
Belleville, Ill.	CST	0.0 "
Peoria, Ill.	CST	– 2.0 "
Rockford, Ill.	CST	– 3.5 "
Springfield, Ill.	CST	– 1.5 "
Province of Cincinnati		
Cincinnati, O.	EST	*38.0 minutes
Cleveland, O.	EST	*26.5 "
Columbus, O.	EST	*32.0 "
Fort Wayne, Ind.	CST	–19.5 "
Indianapolis, Ind.	CST	–15.5 "
Toledo, O.	EST	*34.0 "
Province of Detroit		
Detroit, Mich.	EST	*32.5 minutes
Grand Rapids, Mich.	EST	*42.5 "
Lansing, Mich.	EST	*38.5 "
Marquette, Mich.	EST[1]	*49.5 "
Saginaw, Mich.	EST	*36.0 "
Province of Dubuque		
Dubuque, Ia.	CST	* 2.5 minutes
Cheyenne, Wyo.	MST	– 1.0 "
Davenport, Ia.	CST	* 2.0 "
Des Moines, Ia.	CST	*14.0 "
Grand Island, Neb.	CST	*33.5 "
Lincoln, Neb.	CST	*27.0 "
Omaha, Neb.	CST	*23.5 "
Sioux City, Ia.	CST	*25.5 "
Province of Los Angeles		
Los Angeles, Cal.	PST	– 7.0 minutes
Fresno, Cal.	PST	– 0.5 "
Monterey, Cal.	PST	* 7.5 "
San Diego, Cal.	PST	–11.5 "
Tucson, Ariz.	MST	*24.0 "

[1] EST is the usual time in Marquette, but CST is the legal time according to the Federal law.

Province of Louisville

Louisville, Ky.	CST	–17.0	minutes
Covington, Ky.	CST	–22.0	"
Owensboro, Ky.	CST	–11.5	"
Nashville, Tenn.	CST	–13.0	"

Province of Milwaukee

Milwaukee, Wis.	CST	– 8.5	minutes
Green Bay, Wis.	CST	– 8.0	"
La Crosse, Wis.	CST	* 3.0	"
Superior, Wis.	CST	* 8.5	"

Province of Newark

Newark, N. J.	EST	– 3.5	minutes
Camden, N. J.	EST	* 0.5	"
Paterson, N. J.	EST	– 3.0	"
Trenton, N. J.	EST	– 1.0	"

Province of New Orleans

New Orleans, La.	CST	* 0.5	minutes
Alexandria, La.	CST	*10.0	"
Lafayette, La.	CST	* 8.0	"
Little Rock, Ark.	CST	* 9.0	"
Mobile, Ala.	CST	– 7.5	"
Natchez, Miss.	CST	* 5.5	"

Province of New York

New York, N. Y.	EST	– 4.0	minutes
Albany, N. Y.	EST	– 5.0	"
Brooklyn, N. Y.	EST	– 4.5	"
Buffalo, N. Y.	EST	*15.5	"
Ogdensburg, N. Y.	EST	* 2.0	"
Rochester, N. Y.	EST	*10.5	"
Syracuse, N. Y.	EST	* 4.5	"

Province of Philadelphia

Pittsburg, Pa.	EST	* 1.0	minutes
Altoona, Pa.	EST	*13.5	"
Erie, Pa.	EST	*20.5	"
Harrisburg, Pa.	EST	* 7.5	"
Pittsburgh, Pa.	EST	*20.0	"
Scranton, Pa.	EST	* 2.0	"

Province of Portland in Oregon

Portland, Ore.	PST	*10.5 minutes
Baker City, Ore.	PST	– 8.5 "
Boise, Ida.	MST	*45.0 "
Great Falls, Mont.	MST	*25.0 "
Helena, Mont.	MST	*28.0 "
Seattle, Wash.	PST	* 9.5 "
Spokane, Wash.	PST	–10.5 "
Juneau, Alaska	Alaskan Standard Time[2]	–63.0 "

Province of St. Louis

St. Louis, Mo.	CST	* 1.0 minutes
Concordia, Kan.	CST	*30.5 "
Kansas City, Mo.	CST	*18.5 "
Leavenworth, Kan.	CST	*19.5 "
St. Joseph, Mo.	CST	*19.0 "
Wichita, Kan.	CST	*29.0 "

Province of St. Paul

St. Paul, Minn.	CST	*12.5 minutes
Bismarck, N. Dak.	CST	*43.0 "
Crookston, Minn.	CST	*26.5 "
Duluth, Minn.	CST	* 8.5 "
Fargo, N. Dak.	CST	*27.0 "
Rapid City, S. Dak.	MST	– 7.0 "
St. Cloud, Minn.	CST	*16.5 "
Sioux Falls, S. Dak.	CST	*26.5 "
Winona, Minn.	CST	* 6.5 "

Province of San Antonio

San Antonio, Tex.	CST	*34.0 minutes
Amarillo, Tex.	CST	*47.5 "
Corpus Christi, Tex.	CST	*30.0 "
Dallas, Tex.	CST	*27.0 "
Galveston, Tex.	CST	*19.0 "

[2] Juneau, however, actually uses the time of the 135th meridian and not the legal time of the 150th meridian (Alaskan Standard Time).

Oklahoma City, Okla.	CST	*30.0	"
Tulsa, Okla.	CST	*24.0	"

Province of San Francisco

San Francisco, Cal.	PST	*10.0	minutes
Reno, Nev.	PST	– 0.5	"
Sacramento, Cal.	PST	* 6.0	"
Salt Lake City, Utah	MCT	*27.5	"

Province of Sante Fe

Sante Fe. N. Mex.	MST	* 4.0	minutes
Denver, Col.	MST	0.0	"
El Paso, Tex.	CST	*66.0	"
Gallup, N. Mex.	MST	*15.0	"

TABLE II

TRUE TIME

In this table the asterisk (*) sign will indicate the amount by which mean solar time is ahead of true solar time. The minus (–) sign will indicate the amount by which it is slow according to the variations that occur throughout the year. This table may be used within any one of the Standard Time zones.

Date	Jan.	Feb.	Mar.	Apr.	May	June	July	Aug.	Sept.	Oct.	Nov.	Dec.
1	*4	*14	*12	*4	–3	–2	*4	*6	0	–10	–16	–11
2	*4	*14	*12	*4	–3	–2	*4	*6	0	–11	–16	–11
3	*5	*14	*12	*3	–3	–2	*4	*6	–1	–11	–16	–10
4	*5	*14	*12	*3	–3	–2	*4	*6	–1	–11	–16	–10
5	*6	*14	*12	*3	–3	–2	*4	*6	–1	–12	–16	–9
6	*6	*14	*11	*2	–3	–2	*5	*6	–2	–12	–16	–9
7	*6	*14	*11	*2	–4	–1	*5	*6	–2	–12	–16	–8
8	*7	*14	*11	*2	–4	–1	*5	*6	–2	–12	–16	–8
9	*7	*14	*11	*2	–4	–1	*5	*5	–3	–13	–16	–8
10	*8	*14	*10	*1	–4	–1	*5	*5	–3	–13	–16	–7
11	*8	*14	*10	*1	–4	–1	*5	*5	–3	–13	–16	–7
12	*8	*14	*10	*1	–4	0	*5	*5	–4	–13	–16	–6
13	*9	*14	*10	*1	–4	0	*6	*5	–4	–14	–16	–6

Date	Jan.	Feb.	Mar.	Apr.	May	June	July	Aug.	Sept.	Oct.	Nov.	Dec.
14	*9	*14	*9	0	–4	0	*6	*5	–4	–14	–16	–5
15	*10	*14	*9	0	–4	0	*6	*4	–5	–14	–15	–5
16	*10	*14	*9	0	–4	0	*6	*4	–5	–14	–15	–4
17	*10	*14	*8	0	–4	*1	*6	*4	–5	–15	–15	–4
18	*11	*14	*8	–1	–4	*1	*6	*4	–6	–15	–15	–3
19	*11	*14	*8	–1	–4	*1	*6	*4	–6	–15	–15	–3
20	*11	*14	*8	–1	–4	*1	*6	*3	–7	–15	–14	–2
21	*11	*14	*7	–1	–4	*2	*6	*3	–7	–15	–14	–2
22	*12	*14	*7	–1	–3	*2	*6	*3	–7	–15	–14	–1
23	*12	*14	*7	–2	–3	*2	*6	*3	–8	–16	–14	–1
24	*12	*13	*6	–2	–3	*2	*6	*2	–8	–16	–13	0
25	*12	*13	*6	–2	–3	*2	*6	*2	–8	–16	–13	0
26	*13	*13	*6	–2	–3	*3	*6	*2	–9	–16	–13	*1
27	*13	*13	*5	–2	–3	*3	*6	*1	–9	–16	–12	*1
28	*13	*13	*5	–3	–3	*3	*6	*1	–9	–16	–12	*2
29	*13	*12	*5	–3	–3	*3	*6	*1	–10	–16	–12	*2
30	*13		*5	–3	–3	*3	*6	*1	–10	–16	–11	*3
31	*14		*4		–3		*6	0		–16		*3[3]

Instructions for the Use of the Tables

A—*To obtain the greatest amount of latitude for the observance of the eucharistic fast for instance.*

Consult the two tables. If two positive values are given, add them: the total is the latitude one enjoys after the Standard Time midnight. If the two figures are negative, there is no latitude given and food cannot be taken after midnight of the Standard Time clock. When one figure is positive and the other negative a distinction must be made. If the mean time figure shows a plus quantity, and the true time figure a minus quantity, no matter how large or small the latter may be, then one still retains the full advantage indicated by the plus quantity of the mean time figure. On the other hand, if the mean time figure shows a minus quantity, and the true time figure a plus quantity,

[3] This chart has been taken from the diminutive brochure: *"Advantageous Utilization of the Latitude Offered by Canon 33"*, by the Reverend Sydney Turner, C.P.

then subtract the former from the latter. If the remainder is a plus quantity, it indicates the latitude one enjoys; if the remainder is zero or if it becomes a minus quantity, then no advantage is gained. The use of Standard Time itself will then prove the most advantageous.

For instance, let us figure the true sun midnight in El Paso, Tex. for Feb. 3rd. According to Table I El Paso has a plus quantity of 66 minutes (*66). Feb. 3rd has a plus quantity of 14 minutes. Consequently 66 plus 14 give 80 minutes. The result is that on Feb. 3rd in El Paso the eucharistic fast does not bind before 1:20 A. M. CST, whenever the reckoning is done according to true sun time.[4]

New York City's local mean time is 4 minutes faster than EST (–4 minutes). On the other hand Tabe II gives us *10 minutes for March 10th. Therefore 10–4 leave a latitude of 6 minutes. The deadline would then be 12:06 A. M. EST.

B—*To find the greatest amount of latitude to mark the end of a precept, for instance, that of the law of abstinence on Friday night.*

Simply reverse the preceding operation and try to get the highest negative value.

Thus Boston's mean time is 16 minutes faster than EST (–16 minutes). The difference between mean and true time for Nov. 13th is also –16 minutes. Add the two and you will have the total of 32 minutes. Consequently the law of abstinence ceases on Nov. 13th in Boston at 11:28 P. M. EST, whenever the time is computed according to true solar time.

[4] One must however bear in mind that Mountain Time is the usual time in El Paso, Central Time being the legal time according to the Federal plan.

APPENDIX II

THE CALENDAR REFORM [1]

It might be useful to insert at this point a few words about possible improvements on the Gregorian calendar. For at least a century and a half new calendar reforms have been proposed and even tried. Until recent years, however, these were often prompted by ill feeling toward the Vatican and were really lamentable and pitiable efforts at improvement.

Auguste Comte, for instance, elaborated a 13 month calendar of his own, wherein the weeks were named and the months designated by a new appellation. Thus, the first month was dedicated to Theocratic Civilization and was called "Moses." The week names for that month were: Numa, Buddha, Confucius and Mahomet. Let it suffice to say that plans such as this are not adaptable.[2]

The French Revolution tried a weird calendrical scheme and found to its great sorrow that it did not work. That erratic movement wanted to break away completely from the past. The calendar therefore had to go, whether or not a suitable substitute could be found. The neopagan philosophers turned their eyes to the Egypt of the Pharaohs and adopted a similar calendar in a modified form. This "calendar of reason" contained 365 days, divided into 12 months of 30 days. The last 5 days were epagomenal days similar to those of Egypt. The week was dropped and a 10 day period introduced in its place. Here are the names of the months:

January .... Nivose .... Snowing month
February .... Pluviose .... Rainy month
March .... Ventose .... Windy month

April .... Germinal .... Budding month
May .... Floreal .... Flowery month
June .... Prairial .... Meadows month

[1] Cf. Schwegler, *Catholics and Calendar Reform* (New York: The World Calendar Association, 1934).

[2] Cf. Wilson, *The Romance of the Calendar*, p. 258.

July . . . . Messidor . . . . Harvest month
August . . . . Thermidor . . . . . . Heat month
September . . . . Fructidor . . . . Fruitful month

October . . . . Vendemiare . . . . . Vintage month
November . . . . Brumaire . . . . Foggy month
December . . . . Frimaire . . . . Freezing month.[3]

In 1802, however, Napoleon abolished the 10 day period, and in 1806 restored the Gregorian calendar. In 1871 attempts were made to revive the revolutionary calendar, but they were short-lived.

These so called reformed calendars were nothing but the product of unbridled imaginations. Of late, however, more sensible reforms have been proposed. These in many instances consider themselves the continuation of the work of Gregory XIII. Thus one reads in *The World Calendar:* "To finish what the Gregorian reform left undone is the purpose of the present movement." [4]

After all, the Gregorian calendar was a decisive improvement over its predecessors, but it is not by any means perfect. Some of its disadvantages are rather evident. These may be divided into two groups. There is first of all the fact that Pope Gregory did not attempt to modify the makeshift arrangement of the months and quarters that the Romans more or less improvised. For instance, he left February with its 28 days as he found it. In the second place the shifting of Easter, although directly of liturgical character, constitutes another chronological difficulty.

a) In one way it may be said that the Gregorian calendar is 14 calendars instead of one. There are 14 different ways in which the days of the week can be blended in with the month. Thus January 1 might fall on every day of the week. The leap years simply double the possibilities of variance. Leap years with a uniform arrangement throughout, that is, years in which any given dates within any given months recur on the same week-day, are reproduced only once every 28 years. The calendar we had

[3] Cf. Wilson, *The Romance of the Calendar,* p. 258.
[4] Achelis, The World Calendar (New York: Putnam, 1937), p. 81.

in 1940 will not be used again in the same uniform detail before 1968.

Under the present system it is hard to say on what day of the week events occurred. We all know the date of our birth, but was that day a Friday or a Sunday? Perhaps few of us know. The reason is that the weeks do not fit into the months regularly.

Months furthermore have not a like number of days. These vary in number from 1 to 3 (28-31 days). The various quarters of the year, so important for business transactions, are not equal. The first quarter has 90 days; the second, 91 days; the third, 92 days; and the fourth, 92 days. This difference plays havoc with the precision one would like to find in statistics.

The irregularity of our holidays is also rather annoying, especially to business men.

b) The vagaries of Easter have complicated the problem to a still greater extent. Easter is looked upon as a landmark in the scholastic and in the business world just as in the liturgical. School vacations and examinations, for instance, are based on Easter, but that feast varies from March 22 to April 25. Business men complain that this variance deeply affects the sale of spring and summer clothing. A late Easter will practically stamp out the sale of spring outfits. People will simply pass on from winter to summer wear.

It is no wonder, then, that for some time various associations have been advocating a calendar reform on a rational and utilitarian basis. These associations may be grouped together into two divisions: the advocates of a 13 month calendar and the advocates of a reformed 12 month calendar.

1—The 13 month calendar [5]

The 13 month calendar is proposed by the "International Fixed Calendar League." Its head office is in London; it has a branch office in Rochester, N. Y., where it had one of its strongest supporters in the person of the late George Eastman.

According to this plan the 13 months would all have exactly 4

[5] Cf. Schwegler, *Catholics and Calendar Reform,* p. 12 ff.

weeks and would all begin on a Sunday. The 13th month would be called "Sol" and would be placed between June and July. These 13 months would account for 364 days. The 365th day would fall on December 29 and would be extra-weekly. Every fourth year the leap day would come on June 29. These two days would be international holidays.

The chief advantage of this calendar is its simplicity. It does away with the calendar as we now know it to the extent that a year clock could take its place. The calculation of past and future dates becomes simple.

These advantages are however minimized by the following difficulties:

The change is too radical. It would renew the difficulties of the Old and the New Style.

Number 13 is unfortunate. A quarter would be 3¼ months. Monthly and quarterly accounting would become more complicated.

The last day of the month, which is the biggest one for accounts, would always fall on a Saturday.

All monthly obligations would have to be paid 13 times a year.

Comparative statistics for the past would have to be converted.

Astronomical computations would be complicated.

2—The 12 month calendar

In England, "The Rational Calendar Association," and in the United States, "The World Calendar Association," have been campaigning for an adjustment of the present 12 month calendar. Various other groups have been organized in France, Switzerland, Belgium and Greece.

The World Calendar divides the year into four equal quarters: every year is the same and every quarter is identical. The first month of the quarter has 31 days and the other two, 30 days. The first month always begins on a Sunday, the second on Wednesday, and the third on Friday.

This arrangement gives us 364 days. The extra day would be intercalated after December 30 and would be called "Year-End-Day." It would be an extra-week day or a blank day. "Leap-

Year-Day" would also be a blank day and would follow June 30.

Easter, it is suggested, could fall on April 8 every year. Christmas would always fall on a Monday, thereby giving priests two days to hear the numerous confessions. Ascension Thursday would always be May 16.

If this plan were carried into effect the breviary would be simplified to the extent that the *Proprium de Tempore* and the *Proprium Sanctorum* would be blended together. The "Ordo" would no longer be needed, etc.[6]

The League of Nations has given the matter mature consideration and has sent out questionnaires to the various countries. The reports are the following: Canada and Yugoslavia, and also Portugal, though not definitely, favored a 13 month calendar; Switzerland and Greece advocated a 12 month calendar; Great Britain, Japan, Belgium, the Irish Free State, Holland, Sweden, Germany and a few other countries did not commit themselves, but stated that popular opinion in their territories either favored the 12 month plan or was at least opposed to the 13 month reform; China proposed a calendar of its own; Norway and the United States made no recommendation and showed no preference.

Various religious groups, Protestant and Orthodox, are ready to accept a fixed Easter and a new calendar. Opposition, however, comes from the Jews, the Seventh-Day Adventists and a few other Sabbatarians because the extra-week days would interrupt the succession of Sabbaths.[7]

Catholics should certainly be interested in what the Holy See thinks of all these proposals. Father Schwegler writes:

> Consulted by the League of Nations, the Holy See stated (March 7, 1924), through the Apostolic Nuncio at Berne, that "the question of the reform of the calendar, particularly so far as it concerns the fixing of Easter, is a preeminently religious question and that any changes which might be made in this direction, though they would meet with no difficulty from the point of view of dogma, would nevertheless involve the abandon-

[6] For other proposals cf. Flammarion, *Astronomie Populaire,* p. 30: Philip, *The Calendar,* p. 93 ff.

[7] Cf. Schwegler, *Catholics and Calendar Reform,* pp. 13-14.

> ment of deeply rooted traditions from which it would be neither legitimate nor desirable to depart, except for weighty considerations connected with the general interest." The Holy See further stated that it did not think there was sufficient reason for introducing changes; but that, if some changes were demanded by the public good, the Holy See would not be prepared to consider the question except on the advice of an ecumenical council.[8]

Within the last three or four years the League seems to have lost some of its former enthusiasm.[9]

Of all the proposed calendars the 12 month scheme seems to be the only one with a fair chance of success. The World Calendar ranks with the best of the 12 month reforms. However, past experience has clearly demonstrated that calendar reforms are not as easy to apply as they are to compute. The Gregorian calendar is three and one half centuries old and it is safe to say that it has not been thoroughly accepted by the so called civilized world. "Some nations have adopted the Gregorian Calendar for international purposes, without enforcing it in their internal affairs. At present there are different New Years for Armenians, Moslems, Coptics, Hindus and Hebrews." [10] In 1939 President Roosevelt decided to change the date of Thanksgiving. It is common knowledge that this holiday was observed that year on different days throughout the country.[11]

It must be borne in mind that the good brought about by a new calendar must not be offset by the disadvantages involved in the change. On that basis a reform would not be such a boon to society after all.

It is evident that no change will be made during the present

[8] *Catholics and Calendar Reform,* p. 14.

[9] "The League Council concluded that it is not expedient for the time being to contemplate convening a conference to carry out a reform which in present circumstances would seem to have no chance of being accepted, and that under such conditions, it is unnecessary, until further notice, to retain the question on the agenda."—Schwegler, *ER,* XCVIII (1938), 130.

[10] Anthony M. Turano, "The Calendar is out of Date,"—*The Reader's Digest,"* XXXIV (1939), No. 205, p. 95.

[11] Cf. *The World Almanac,* 1940, p. 120.

war. A Sunday will however begin the year in 1950. That year would be at least theoretically adapted to the introduction of a calendar such as the World Calendar.

The introduction of a 12 month calendar would have practically no effect on the reckoning of time in Canon Law. Nothing would be changed. At most, provision might be made for the computation of the extra-week days. A 13 month calendar would have months of 28 days. The definition of the month as found in canon 32, § 2, might then advantageously be modified.[12]

[12] For more details concerning the reform of the calendar in general cf. also McNally, "The Problem of the Calendar," —*HPR*, XXX (1930), 1033-1041; Schwegler, "The Vatican and a Fixed Easter," —*HPR*, XXXIV (1934), 704-714, and "The Vatican and Calendar Reform," —*HPR*, XXXIV (1934), 801-808; Puig, "Los Proyectos de Reforma del Calendario,"—*Razón y Fe*, XCI (1930), 4-17, 149-156; the *Journal of Calendar Reform*, etc.

BIBLIOGRAPHY

Sources

Acta Apostolicae Sedis, Commentarium Officiale, Romae, 1909-

Acta Sanctae Sedis, 41 vols., Romae, 1865-1908.

Bullarium Diplomatum et Privilegiorum Sanctorum Romanorum Pontificum, Taurinensis Editio, 24 vols. et Appendix, Augustae Taurinorum-Neapoli, 1857-1872.

Codex Iuris Canonici Pii X Pontificis Maximi iussu digestus, Benedicti Papae XV auctoritate promulgatus, Romae: Typis Polyglottis Vaticanis, 1917.

Codicis Iuris Canonici Fontes cura Emi Petri Card. Gasparri editi, 9 vols., Romae (later Civitate Vaticana): Typis Polyglottis Vaticanis, 1923-1939. (Vol. VII-IX *ed. cura et studio Emi Iustiniani Card. Serédi.*)

Corpus Iuris Canonici, ed. Lipsiensis 2., Aemilius Ludovicus Richter-Aemilius Friedberg, 2 vols., Lipsiae, 1879-1881.

Corpus Iuris Civilis (Kreuger-Mommsen-Scholl-Kroel), 5. ed., 3 vols., Berolini, 1928-1929.

Corpus Juris Secundum, A Complete Restatement of the Entire American Law as developed by all Reported Cases, William Mack-Donald J. Kiser, 19 vols., Brooklyn: The American Law Book Co., 1936-

Decreta Authentica Congregationis Sacrorum Rituum ex actis eiusdem collecta cura et studio Aloisii Gardellini, 3. ed., 5 vols., Romae, 1856-1879.

Decreta Authentica Congregationis Sacrorum Rituum ex actis eiusdem collecta eiusque auctoritate promulgata, 7 vols., Romae, 1898-1927.

Decretales D. Gregorii Papae IX, una cum Glossis Restitutae, Romae, 1582.

Decretum Gratiani Emendatum et Notationibus Illustratum, una cum Glossis, 2 vols., Romae, 1582.

Holy Bible, The, Translated from the Latin Vulgate (Douay Version), Baltimore, 1914.

Jaffé, Philippus, *Regesta Pontificum Romanorum ab condita Ecclesia ad annum post Christum natum MCXCVIII,* 2 vols. in 1, Lipsiae, 1881-1888.

Liber Sextus Decretalium D. Bonifacii Papae VIII, suae integritati una cum Clementinis et Extravagantibus, earumque Glossis Restitutus, Romae, 1582.

Migne, Jacques Paul, *Patrologiae Cursus Completus, Series Latina,* 221 vols., Parisiis, 1844-1864.

Sacri Concilii Tridentini Canones et Decreta, Coloniae Agrippinae, 1621.

United States Code Annotated, 65 vols., St. Paul: West Publishing Company, 1927-

United States Reports, Cases Adjudged in the Supreme Court, 307 vols., Washington: United States Government Printing Office.

AUTHORS

Achelis, Elisabeth, *The World Calendar,* New York: Putnam, 1937.

Alphonsus, St., *Theologia Moralis,* 9 vols., Vesuntione, 1828.

D'Angelo, Sosius, *Ius Digestorum,* 2 vols., Romae: Athenaeum Pontificii Seminarii Romani ad "S. Apollinaris," 1927-1928.

D'Annibale, Josephus, *Summula Theologiae Moralis,* 2. ed., 3 vols., Mediolani, 1881-1883.

Antonellus, Ioannes, *Tractatus Novissimus et Absolutissimus de Tempore Legali,* Venetiis, 1692.

Ayrinhac, H. A., *Marriage Legislation in the New Code of Canon Law,* Revised, Enlarged by P. J. Lydon, New York: Benziger, 1932.

(Bachofen), Charles Augustine, *A Commentary on the New Code of Canon Law,* 8 vols., Vol. I, 3. ed., St. Louis: Herder, 1920.

Baldus de Ubaldis, *In Decretalium Volumen Commentaria,* Venetiis, 1580.

Bakalarczyk, Richardus, *De Novitiatu,* The Catholic University of America, Canon Law Studies, n. 36, Washington: The Catholic University of America, 1927.

Barbosa, Augustinus, *Tractatus Varii,* Lugduni, 1660.

Bastien, Pierre, *Directoire Canonique à l'Usage des Congrégations à Voeux Simples,* 3. ed., Bruges: Charles Beyaert, 1923.

Benedictus XIV, *Opera Omnia,* 17 vols. in 18, Prati, 1739-1747.

Beste, Ulricus, *Introductio in Codicem,* Collegeville, Minn.: St. John's Abbey Press, 1938.

Blat, Albertus, *Commentarium Texus Codis Iuris Canonici,* 6 vols., Romae, 1921-1927.

Bonacina, Martinus, *Opera Omnia,* 3 vols., Venetiis, 1687.

Bouscaren, T. Lincoln, *The Canon Law Digest,* 2 vols., and Supplement—1938, Milwaukee: Bruce, 1934-1938.

Cabrol, Fernand, *Dictionnaire d'Archéologie Chrétienne et de Liturgie,* 14 vols. in 27, Paris: Letouzey et Ané, 1924-

Calà, Caesar, *Tractatus Absolutissimus de Feriis, Solemnibus, Repentinis, et Indictis,* Neapoli, 1675.

Calmet-James-Migne, *Dictionnaire Historique, Archéologique, Philologique et Littéral de la Bible,* 4. ed., 4 vols. in 2, Paris, 1846.

Cance, Adrien, *Le Code de Droit Canonique,* 3 vols., Paris: Gabalda, 1927-1929.

Cappelli, A., *Cronologia, Cronografia e Calendario Perpetuo,* 2. ed., Milano: Ulrico Hoepli, 1930.

Cappello, Felix M., *Summa Iuris Canonici,* 3 vols., Romae: Apud Aedes Univ. Greg., Vol. I, 2. ed., 1932.

———, *Tractatus Canonico-Moralis de Sacramentis,* Vol. I, *De Sacra-*

mentis in Genere, de Baptismo, Confirmatione et Eucharistia, Taurinorum Augustae: Marietti, 1921.

———, *Tractatus Canonico-Moralis de Sacramentis,* Vol. I, *De Sacramentis in Genere, de Baptismo, Confirmatione et Eucharistia,* 3. ed., Taurinorum Augustae: Marietti, 1938.

Catholic Encyclopedia, The, 15 vols., New York, 1907-1912.

Catholic Encyclopoedic Dictionary, The, London: Cassell, 1931.

Chelodi, Ioannes, *Ius de Personis,* 2. ed. ab Ernesto Bertagnolli, Tridenti: Ardesi, 1927.

Cicognani, Amleto, *Canon Law,* Authorized English Version by J. O'Hara and F. Brennan, Philadelphia: Dolphin Press, 1934.

Claeys-Boúúaert, *Selecta Capita Codicis Iuris Canonici,* Gandae, 1919.

Cocchi, Guidus, *Commentarium in Codicem Iuris Canonici ad Usum Scholarum,* 5 vols. in 8, Taurinorum Augustae: Marietti, Vol. I, 2. ed., 1921.

Coronata, Matthaeus Conte A, *Institutiones Iuris Canonici,* 5 vols., Taurini: Marietti, 1928-1936.

Covarruvias, Didacus, *Opera Omnia,* 2 vols., Coloniae Allobrogum, 1679.

De Angelis, Philippus, *Praelectiones Iuris Canonici ad Methodum Decretalium Gregorii IX Exactae,* 4 vols. in 6, Romae, 1877-1887.

De Becker, Iulius, *De Matrimonio Praelectiones Canonicae,* ed. nova ad tramites Codicis Iuris Canonici accomodata, Louvain: Fr. Ceuterick, 1931.

Decius, Philippus, *In Decretales Commentaria,* Lugduni, 1576.

De Justis, Vincentius, *De Dispensationibus Matrimonialibus,* Lucae, 1727.

De Lugo, Ioannes, *Disputationes Scholasticae et Morales,* ed. nova, 8 vols., Parisiis, 1868-1869.

De Meester, Alphonsus, *Iuris Canonici et Iuris Canonico-Civilis Compendium.,* 3 vols. in 4, Brugis: Sumptibus et Typis Societatis Sancti Augustini, 1921-1928.

Donelli, Hugo, *Opera Omnia Commentariorum de Iure Civili, cum Notis Osualdi Hilligeri,* 12 vols., Romae, 1828.

Du Cange (Carolus Dufresne), *Glossarium ad Scriptores Mediae et Infimae Latinitatis,* ed. nova, 6 vols., Parisiis, 1733-1736.

Eichmann, Eduard, *Lehrbuch des Kirchenrechts auf Grund des Codex Iuris Canonici,* 2. ed., Paderborn: Druck und Verlag von Ferdinand Schöningh, 1926.

Encyclopaedia Britannica, The, 14. ed., 24 vols., London: The Encyclopaedia Britannica Co., 1936.

Encyclopedia Americana, The, 30 vols., New York: Americana Corporation, 1938.

Engel, Ludovicus, *Collegium Universi Iuris Canonici,* 9. ed., Beneventi, 1760.

Falco, Mario, *Introduzione allo Studio del "Codex Iuris Canonici,"* Torino: Fratelli Bocca, 1925.

Fanfani, Ludovicus, *De Iure Religiosorum,* 2. ed., Taurini: Marietti, 1925.

Ferraris, Lucius, *Bibliotheca Canonica, Juridica, Moralis, Theologica necnon Ascetica, Polemica, Rubricista, Historica,* 9 vols., Romae, 1885-1899. (Vol. IX ed. Ianuarius Bucceroni.)

Flammarion, Camille, *Astronomie Populaire,* Paris, 1890.

Gasparri, Petrus, *Tractatus Canonicus de Matrimonio,* 2 vols., Parisiis, 1891.

Gavantus, Bertholomeus, *Thesaurus Sacrorum Rituum seu Commentaria in Rubricas Missalis et Breviarii Romani,* ed. novissima, 2 vols., Coloniae Agrippinae, 1736.

Gennari, Casimir, *Consultations de Morale, de Droit Canonique et de Liturgie,* Authorized Translation from the Italian Text by A. Boudinhon, 5 vols., Paris, 1907-1910.

Godfray, Hugh, *A Treatise on Astronomy,* 4. ed., London, 1886.

Gonzalez-Tellez, Emmanuel, *Commentaria Perpetua in Singulos Textus Quinque Librorum Decretalium Gregorii IX,* 4 vols., Venetiis, 1699.

Hefele, Carolus, et Leclercq, Henricus, *Histoire des Conciles d'après les Documents Originaux,* 10 vols. in 19, Paris, 1907-1938.

Haring, Johann B., *Grundzüge des katholischen Kirchenrechtes,* 3. ed., 2 vols., Graz: Ulrich Mosers Buchhandlung, 1924.

Herrera, Antonio Parra, *Legislacion Eclesiastica sobra el Ayuno y la Abstinencia,* The Catholic University of America, Canon Law Studies, n. 92, Washington: The Catholic University of America, 1935.

Hilling, N., *Die allgemeinen Normen des Codez Iuris Canonici,* Freiburg i. Br.: Verlag von Josef Waibel, 1926.

Jorio, Domenico, *La Comunione agl'Infermi,* Romae: Pustet, 1931.

Kealy, John James, *The Introductory Libellus in Church Court Procedure,* The Catholic University of America, Canon Law Studies, n. 108, Washington: The Catholic University of America, 1937.

Lacau, Ioannes, *De Tempore, Dissertatio Philosophico-Scientifico-Iuridica,* Augustae Taurinorum: Marietti, 1921.

La Croix, Claudius, *Theologia Moralis,* Coloniae, 1739.

Laymann, Paul, *Theologia Moralis,* 6. ed., 2 vols., Bambergae, 1669.

Leage, R. W.-Ziegler, C. H., *Roman Private Law,* 2. ed., London: MacMillan, 1937.

Lega, Michael, *Praelectiones in Textum Iuris Canonici de Delictis et Poenis,* 2. ed., Romae, 1910.

———, *Praelectiones in Textum Iuris Canonici de Iudiciis Ecclesiasticis,* 4 vols., Romae, 1896-1901.

Lehmkuhl, Augustinus, *Theologia Moralis,* 12. ed., 2 vols., Friburgi Brisgoviae: Herder, 1914.

Leitner, Martin, *Handbuch des katholischen Kirchenrechts auf Grund des neuen Kodex,* 5 vols., Vol. I, 2. ed., Regensburg: Pustet, 1921.

Many, S., *Praelectiones de Missa,* Paris, 1903.

Marc-Gestermann-Raus, *Institutiones Morales Alphonsianae,* 18. ed.. 2 vols., Lugduni: Vitte, 1927.

Maroto, Philippus, *Institutiones Iuris Canonici ad Normam Novi Codicis*, 2 vols., Vol. I, 3. ed., Romae: Apud Commentarium pro Religiosis, 1921.

Maynz, Charles, *Cours de Droit Romain*, 3. ed., 3 vols., Bruxelles, 1870.

McKenna, Stephen, *Paganism and Pagan Survivals in Spain up to the Fall of the Visigothic Kingdom*, The Catholic University of America. Studies in Medieval History, New Series. Vol. I, Washington: The Catholic University of America, 1938.

Medicus, *Medical Essays*, 3. ed., 1931.

Michiels, Gommarus, *Normae Generales Juris Canonici*, 2 vols., Lublin: Universitas Catholica, 1929.

Newcomb-Holden, *Astronomy*, 3. ed., New York, 1887.

Nys, D., *Cours de Philosophie*, Vol. VII, *La Notion de Temps*, 3. ed., Tome III, Louvain: Institut de Philosophie, 1925.

Oesterle, Gerardus, *Praelectiones Iuris Canonici*, Vol. I, Romae: In Collegio S. Anselmi, 1931.

Ojetti, Benedictus, *Commentarium in Codicem Iuris Canonici*, 4 vols., Romae: Apud Aedes Univ. Greg., 1927-1931.

————, *Synopsis Rerum Moralium et Iuris Pontificii*, Romae, 1899.

Packer, George, *Our Calendar*, Corning, N. Y., 1893.

Panormitanus, Abbas (Nichalaus de Tudeschis), *Commentaria in Quinque Libros Decretalium*, 5 vols., in 7, Venetiis, 1588.

Pastor, Ludwig Freiherr von, *The History of the Popes From the Close of the Middle Ages*, 29 vols., translation, Vols. I-VI, ed. by Frederick I. Antrobus; Vols. VII-XXIV, ed. by Ralph F. Kerr; Vols. XXV-XXIX, ed. by Dom Ernest Graf, St. Louis: Herder, 1906-1938.

Pejska, Josephus, *Ius Canonicum Religiosorum*, 3. ed., Friburgi Brisgoviae: Herder, 1927.

Philip, Alexander, *The Calendar*, Cambridge: Cambridge University Press, 1921.

Phillips, R. P., *Modern Thomistic Philosophy*, 2. ed., 2 vols., Vol. I, London: Burns Oates and Washbourne, 1939.

Pirhing, Enricus, *Jus Canonicum in V Libros Decretalium*, 4 vols., Dilingae, 1722.

Raus, Joannes B., *Institutiones Canonicae*, 2. ed., Lugduni: Vitte, 1931.

Reiffenstuel, Anacletus, *Ius Canonicum Universum*, 4 vols., Venetiis, 1735.

Roberti, Franciscus, *De Processibus*, 2 vols., Romae: Apud Aedes Facultatis Iuridicae ad S. Apollinaris, 1926.

Sabetti, Aloysius, *Compendium Theologiae Moralis*, 3. ed., New York, 1888.

Sanchez, Thomas, *De Sancto Matrimonii Sacramento*, 3 vols., Antverpiae, 1626.

Schäfer, Timotheus, *Compendium de Religiosis ad Normam Codicis Iuris Canonici*, 2.ed., Münster i. W.: Aschendorffsche Verlagsbuchhandlung, 1931.

Schmalzgrueber, Franciscus, *Ius Ecclesiasticum Universum*, 12 vols., Romae, 1843-1845.

Schwegler, Edward, *Catholics and Calendar Reform*, New York: The World Calendar Association, 1934.

Standard Time (Excerpt from Interstate Commerce Commission Activities 1887-1937), Washington, 1937.

Standard Time Throughout the World (Circular of the National Bureau of Standards C406), Washington: United States Government Printing Office, 1935.

Suarez, Franciscus, *Opera Omnia*, 28 vols., Parisiis, 1856-1878.

Tagle, Ernesto Gomez, *De Temporis Supputatione* (Typewritten Manuscript), Washington, 1932.

Tanquerey, Ad., *Synopsis Theologiae Moralis et Pastoralis*, Tomus III, 6. ed., Romae: Desclée, 1921.

Tiraquellus, Andreas, *Commentaria de Utroque Retractu, et Municipali, et Conventionali*, 4 ed., Venetiis, 1562.

Toso, Albertus, *Ad Codicem Iuris Canonici Commentaria Minora*, Vol. I, Tiferni Tiberini: Typographia Vinciana, 1921.

Van Hove, A., *Commentarium Lovaniense in Codicem Iuris Canonici*, Vol. I, Tomus II, *De Legibus Ecclesiasticis*, Mechliniae: Dessain, 1930; Vol. I, Tomus III, *De Consuetudine-De Temporis Supputatione*, Mechliniae: Dessain, 1933.

Vermeersch, Arthurus, *Theologiae Moralis Principia, Responsa, Consilia*, 2. ed., 3 vols., Romae: Univ. Greg., 1926.

Vermeersch, A., et Creusen, J., *Epitome Iuris Canonici*, 3 vols., Vols. I et II, 5. ed., Vol. III, 3. ed., Mechliniae: Dessain, 1928-1934.

Vromant, G., *Ius Missionariorum*, Vol. I, *Introductio et Normae Generales*, Louvain: Museum Lessianum, 1934.

Wernz, Franciscus X., *Ius Decretalium ad Usum Praelectionum in Scholis Textus Canonici sive Iuris Decretalium*, 6 vols., Romae, 1899-1904.

Wernz, Franciscus, et Vidal, Petrus, *Ius Canonicum*, 7 toms. in 9 vols., Romae: Apud Aedes Universitatis Gregorianae, 1923-1938.

Wilson, P. W., *The Romance of the Calendar*, New York: Norton, 1937.

Windscheid, Bernard, *Diritto delle Pandette*, Traduzione dei professori Carlo Fadda e Paolo Emilio Bensa, 5 vols., Torino: Unione Tipografico-Editrice Torinese, 1925-1926.

World Almanac, The, and Book of Facts, 1940, New York World Telegram.

Woywod, Stanislaus, *A Practical Commentary of the New Code of Canon Law*, 4. ed., 2 vols., New York: Wagner, 1932.

———, *The New Canon Law, A Commentary and Summary of the New Code of Canon Law*, new ed., New York: Wagner, 1918.

Young, Charles A., *Elements of Astronomy*, revised ed., Boston, 1903.

Periodicals

American Historical Review, The, New York, 1895-
L'Ami du Clergé, Paris, 1878-
Apollinaris, Romae, 1928-
Archiv für katholisches Kirchenrecht, Innsbruck, 1857-1861; Mainz,1862-
Canoniste Contemporain, Le, Paris, 1878-1926.
Commentarium pro Religiosis (later *Commentarium pro Religiosis et Missionariis*), Romae, 1920-
Ecclesiastical Review, The (originally *The American Ecclesiastical Review*), Philadelphia, 1889-
El Monte Carmelo, Burgos, 1900-
Homiletic and Pastoral Review, The, New York, 1900-
Irish Ecclesiastical Record, Dublin, 1864-
Journal of Calendar Reform, New York, 1931-
Jus Pontificium, Romae, 1921-
Monitore Ecclesiastico, Il, Romae, 1876-
Nederlandsche katholieke Stemmen, Zwolle, 1901-
Nouvelle Revue Théologique, Paris, 1869-
Pastor Bonus, Trier, 1889-
Periodica de Re Canonica et Morali utili praesertim Religiosis et Missionariis, Brugis, 1905-; ab anno 1927: *Periodica de Re Morali, Canonica, Liturgica.*
Razón y Fe, Madrid, 1900-
Revue Ecclésiastique de Liége, Liége, 1909-
Theologisch-praktische Quartalschrift, Linz, 1832-

Articles

D'Angelo, Sosius, "Decadentia,"—*Apollinaris,* II (1929), 36-52.
Browne, M. J., "Zonal Time,"—*IER,* XLIII (1934), 420-422; 645.
Cardenas, Enrique M., "Midnight and Canon Law,"—*ER,* CI (1939), 399-413.
Collins, John J., "Can the Star of the Magi Give us the Date of Christ's Birth,"—*ER,* CI (1939), 551-555.
Creusen, J., "Minuit Canonique" ou "Loi Pure et Simple,"—*NRT,* L (1923), 464-474.
Fallon, M. J., "Do Differences in Calendars Affect the Obligation of the 'Missa pro Populo',"—*IER,* XXXVII (1931), 638-641.
Ferland, J., "L'Avance de l'Heure et Certains Préceptes de l'Eglise,'—*Semaine Religieuse de Québec,* XXV (1922). 280-283; 294-298; 567-570; 600-605; 614-621.
Fleming, Sanford, "Time-Reckoning for the 20th Century,"—*Annual Report of the Board of Regents of the Smithsonian Institution,* 1886, Part I, 345-366.

Gerard, John, "Chronology,"—*CE*, III, 738-742.

Hecht, F. X., "Die Berechnung des Schalttages in Noviziatsjahr,"—*AKKR*, CIV (1924), 278-282;—*LQS*, LXXVIII (1925), 149-152.

Jombart, E., "La Ligne de Démarcation de Date,"—*NRT*, LV (1928), 768.

Kieffer, G., "Uber Anwendung zweier verschiedener Probabilitäten und Zeitsysteme,"—*Pastor Bonus*, XL (1929), 286-295.

Kinane, J., "Some Queries in Regard to the Modification of the Eucharistic Fast for the Sick,"—*IER*, XXXV (1930), 519-521.

Lamont, Roscoe, "Augustan Myths,"—*Journal of Calendar Reform*, IX (1939), 10-13.

Leclercq, H., "Les Jours de la Semaine,"—*Dictionnaire d'Archéologie Chrétienne et de Liturgie*, VII, 2736-2745.

Leroux, E., "L'Heure à Suivre dans l'Observation des Lois Ecclésiastiques,"—*Revue Ecclésiastique de Liège*. XI (1919-1920), 157-164.

McNally, Paul A., "The Problem of the Calendar,"—*HPR*, XXX (1930), 1033-1041.

Mitchell, Hugh C., "True Sun Midnight by Your Watch for 1934,"—*ER*, XC (1934), 77-83.

———, "What Time is it? Midnight and Fasting,"—*ER*, LXXXIV (1931), 491-500.

Niebecker, Engelbert, "Die Berechnung der Tagesstunden (Zeitpunkt der Mitternacht usw.) nach dem Can. 33 des C. I. C.,"—*LQS*, LXXXIII (1930), 763-776.

Oesterle, G., "Die Berechnung des Schalttages für das Noviziatsjahr,"—*AKKR*, CIII (1923), 148-149.

———, "Die Dauer des Noviziates,"—*LQS*, LXXIII (1920), 420-424.

Phillips, Edmund S., "The Year and its Parts,"—*ER*, XCII (1935), 1-13.

Puig, E., "Los Proyectos de Reforma del Calendario,"—*Razón y Fe*, XCI (1930), 4-17; 149-156.

Riegel, Robert E., "Standard Time in the United States,"—*The American Historical Review*. XXXIII (1927-1928), 84-89.

Salsmans, J., "De l'Usage Simultané de la Double Probabilité,"—*NRT*, IL (1922), 148-150.

Schwegler, Edward S., "A New Calendar by 1939,"—*ER*, XCVIII (1938). 125-131.

———, "The Vatican and a Fixed Calendar,"—*HPR*, XXXIV (1934), 704-714.

———, "The Vatican and Calendar Reform,"—*HPR*, XXXIV (1934), 801-808.

———, "Sun Time Simplified,"—*ER*, CI (1939), 1-8.

Teodori, I. "De Temporis Supputatione,"—*Apollinaris*, III (1930), 617-619.

Twomey, Louis J., "The Eucharistic Fast,"—*ER*, CII (1940), 405-430.

Van de Loo, F., "Greenwich, M(idden) E(uropesche) of plaatselijke tijd,"—*Nederlandsche katholieke Stemmen*, XXIX (1929), 100-111.

Vermeersch, Arthurus, "A quonam momento estne Professo a votis temporariis integrum religionem deserere ad normam can. 637?,"—*Periodica,* XXII (1933), 34*-39*.

————, "De Non Nullis Supputationibus ad Usum Religiosorum,"—*Periodica,* XVII (1928), 80*-82*.

————, "De Ratione Anni Bissextilis Habenda in Recta Temporis Computatione,"—*Periodica,* I (1911), 3-5.

————, "Recta computatio mensis, quo elapso, licet semel vel bis in hebdomada, decumbenti infirmo permitere ut s. dape reficiatur postquam aliquid per modum potus vel medicinae sumpserit, (can. 858, § 2.,"—*Periodica,* XXIII (1934), 61*-63*.

————,"Supputatio Anni Novitiatus,"—*Periodica,* XVII (1928), 49*.

Voltas, "De Novitiatus Interruptione,"—*Commentarium pro Religiosis,* II (1921), 76-85.

Weigert, G., "Ieiunium naturale und die drei Weinachtsmessen,"—*LQS,* LXXX (1927), 335-336.

Abbreviations

AAS—Acta Apostolicae Sedis.
AKKR—Archiv für katholisches Kirchenrecht.
ASS—Acta Sanctae Sedis.
C.—*Codex* (Iustinianus).
CE—Catholic Encyclopedia.
C.J.S.—*Corpus Juris Secundum.*
CLD—Canon Law Digest (Bouscaren).
CST—Central Standard Time.
D.—*Digestum* (Iustinianum).
DST—Daylight Saving Time.
ER—Ecclesiastical Review.
EST—Eastern Standard Time.
Fontes—Codicis Iuris Canonici Fontes cura . . . Gasparri editi.
HPR—Homiletic and Pastoral Review.
IDL—International Date Line.
IER—Irish Ecclesiastical Record.
LQS—Theologisch-praktische Quartalschrift.
MPL—Migne, *Patrologia Latina.*
MST—Mountain Standard Time.
Nov.—*Novella* (Iustiniana).
NRT—Nouvelle Revue Théologique.
Periodica—Periodica de Re Canonica et Morali Utili praesertim Religiosis et Missionariis.
PCI—Pontifical Commission for the Authentic Interpretation of the Canons of the Code.
PST—Pacific Standard Time.
U.S.C.A.—*United States Code Annotated.*

BIOGRAPHICAL NOTE

Arthur J. Dubé was born on July 22, 1911, in South Berwick, Me., where he attended the St. Rose parochial school. His secondary education was received at Assumption College, Worcester, Mass. He obtained the degree of Bachelor of Arts from this institution in 1933. After four years of theological studies pursued at the Holy Heart Seminary, Halifax, N. S., he was ordained to the priesthood on May 22, 1937. During the next two years he was enrolled as a student at the Pontifical Lateran Athenaeum, Rome, where he received the degree of Baccalaureate in Canon Law in 1938, and the degree of Licentiate in Canon Law in 1939. In September of that year he entered the School of Canon Law at the Catholic University of America.

ALPHABETICAL INDEX

Accuracy in dates, 22
Acts of the same nature,
  actual renewal required, 228
  notion, 224
  various reckonings proposed, 225 ff.
Age,
  cessation of fasting, 252
  Confirmation, 245
  episcopate, pre-Code, 86
  marriage, pre-Code, 84-85
    reckoning in solar years only, 197
  novitiate, pre-Code, 85
  puberty, 246
  sponsors 253
  subdiaconate, pre-Code, 81
  tonsure, pre-Code, 86
Alaska, repetition of one day, 49
Aphelion, 32
Appeals, 237
Arctic circle, 94
Available time,
  evolution of, 67
  exceptional,
    Canon Law, 103, 240
    Roman law, 68
  for impediments of short duration only, 68
  granting of, Code, 240
  impediments to flow of, Code, 230
  lack of new provisions in the Code, 230
  mitigated form of, Code, 243
  notion of, 66, 230
  obstacles to flow of, Roman law,
    moral nature of, 69
    various kinds, 68
  occurrence of, Code, 240
  reckoning of, Code, 234
  Roman law, 66
  starting point always determined, Code, 234
  various forms of, Roman law, 69
Ave Maria bell, 40

Benefices, theory of Antonellus, 84
Bissextilis dies, 16
Breviary,
  choral recitation considered public, 135
  private recitation explained, 136

Calendar,
  Babylonian, 9
  determination of, pre-Code, 77
  Egyptian, 8
  French Revolutionary, 265
  Greek, 9
  Jewish, 52 ff.
  lunar, 7
    use in Canon Law, 78
  luni-solar, 9
  Mahometan, 8
  Roman,
    Augustan changes, 16
    Julian reform, 15
    old Roman, 11
    Roman monthly calendar, 14
  solar, 8
  thirteen month, 267
  World calendar, 266
Calendar reckoning, Code
  ecclesiastical calendar, 192, 198
  Gregorian or lunar calendar, 195
  instances prescribed, 191
  nature of, 191 ff.
  scholastic calendar, 197-198
Calendar reform,
  chance of success, 270
  Holy See, 269
  League of Nations, 269
  pre-Gregorian, 18
Calendars, number and variety of, 10
Canonical equity and vacations, 128

Canonical provision of vacant ecclesiastical offices, 240
Christian era,
  adoption of, 28-29
  date of Christ's birth, 29
Civil law,
  contracts, 116
  meaning of the term, 120
  special computation of month and day, 117
Civil reckoning,
  determination of the first day, 218
  nature of, 61, 217
  occurrence of,
    Canon Law, 80, 217 ff.
    Roman law, 61
Clock,
  celestial, 6
  strokes of, 101, 152-153
Clocks,
  confusion of, 47
  problems when erroneously divergent, 154
Code,
  German, 106
  immediate sources of, 105
Completion of time,
  pre-Code, 80, 83
  Roman law, 66
Continuous time, 210
  de facto, 211
  de iure, 211
  Roman law, 67
Contracts, 116 ff.
  American civil law, 116
  binding force of civil law, 119
  extension of special reckoning, 118
  pre-Code, 105
Creation, calendrical import of, 26
Criminal action, prescription of, 242
Custom of the place,
  calendar, 196
  midnight to midnight reckoning, 91
  usual time, 144

Day,
  beginning of in canon 34, § 3, 219
  canonical duration of, 125
  canonical elements of, 127
  concordance of Roman and modern day, 39
  determining factor of, 38
  fractions of, 128
  indivisible unit, 126
  Jewish, 40
  lunar, 34
  modern, concept of, 41
  Moslem, 40
  notable part of, 129
    absence from novitiate, 209
    available time, 238
  odious affairs, 90
  primitive, duration of, 38
  Roman legal, 58
  sidereal, 42
  solar, 43
Daylight Saving Time,
  canonical status of, 148
  extraordinary legal time, 148-149
  national legal time, 150
  usual time, 145
  where legal in the United States, 150
Days,
  epagomenal, 8
  vacation, intermittent reckoning of, 128
Dawn,
  definition of, 101
  guiding norm when there is no actual dawn, 102
  oneness of reckoning, 132
Dies certa and similar expressions, 104
Dies intercaluris, Roman law, 65
Documents, papal,

mode of dating, 29
Double probability,
application to time-reckoning, 153
distinct obligations, 157-158
relative to eucharistic fast, 157
sinful in general, 154-155

Easter,
determination of, 55-57
vagaries of, 267
Eccentricity of the earth's orbit, 44
Ecclesiastical offices, vacant,
appointments to, 237
Elections, 241
Ember days, successive,
continuity in reckoning two successive ember days, 169 ff.
import of canon 35, 174
lawfulness of separation, 174
Epikeia and the divine office, 178
Episcopal consecration,
required age, pre-Code, 81
Eras, list of, 26
Eucharistic fast, 137
Evening meal as a starting point, 92
Exceptions to the general norms,
ab homine, 114
a iure, 114
instances in the Code, 115
implicit, 121
Exhaustive enumeration in canon 33, § 1,
arguments in favor of, 139
Fast and abstinence,
private observance of, 138
separation of on the same day, 166

Fast and abstinence days,
reckoning of, 129
Favorabilia,
canonical import, 72
first stroke of the clock, 153
leap days, 74
natural reckoning, pre-Code, 79
terminus ad quem, pre-Code, 84

General norms
extension of, 123
prior to Code, 72
Golden number, 10
Greenwich, meridian of, 48
Gregorian,
calendar,
disadvantages of, 266
Eastern churches, 23
evolution of, 11 ff.
innovations of, 19
promulgation of, 19
time of acceptation, 20 ff.
unknown in some regions, 196
reform, 18
year, length of, 20

Holy Communion,
time of distribution, 136
Hour, end of, 152
Hours,
absence from canonical hours, 126-127
multiple reckoning of, number of institutes involved, 138

Identical date, day of, 220
apparent contradiction in the Code, 221
proposed solution, 222
writer's opinion, 222
deficiency of, 221
dissimilarity with German Code, 223
Ides, 13
Ignorance,
available time, 230-231
Roman law, 69
Ignorance and corruption,
Roman calendar, 12
Illustrative enumeration in canon canon 33, § 1,
confirmation of the Sacred Penitentiary, 141
practical use of, 142

reasons for, 140, 142
rescript of Aug. 9, 1899, 141
Impediments in available time,
nature of, 231
notion of duration, 237
work, 239
Imperfections, Roman calendar, 15
Indictions, 28
Indulgences,
visitation of churches, 138
Inferior legislators,
powers relative to time-reckoning, 123
Intention of using a certain reckoning, 176
Intercalations,
Jewish, 54
Mercedonius, 12
Roman calendar, 11, 15
Intermittent time,
various species, 210
International Date Line, 49
practical norms, 186-188

Judges,
determining a time-reckoning, 124
Julian calendar,
advantages of, 18
fundamental error, 18

Kalends, 13
Knowledge of time-units,
determining factor, 6
various phenomena, 7

Laws,
exceptions to, 140
importance of the text, 163
Leap days,
Augustus, 16
Caesar, 15
Leap year,
duration of, pre-Code, 73
length of in Roman law, 66
occurrence of in Gregorian calendar, 19
Legal time, extraordinary, 148
Legitimacy, 249
Liturgical,
books, 112
laws, nature of, 111-112
Locus, elastic meaning, of 145
Lunations and solar years, 9

Mass,
celebration,
private, 133-134
public, 133, 134
time of, 131
exceptions to the general rule, 131
Mathematical reckoning, 88
Mean solar time, 43
time of introduction, 46
Metonic cycle, 9
Midnight,
canonical, 159
solar, 42
Midnight Masses, change of reckoning not licit, 177
Midnight to midnight reckoning,
exclusive use of, 93
Missa pro populo and the International Date Line, 187
Moon,
importance as time-keeper, 52
Neomenia, 53
phases of, 34
Month,
arbitrary unit today, 33
calendrical, various kinds, 33
canonical,
Code, 130
pre-Code, 76-77
Greek, 10
lunar, 53
reckoning of for the sick, 249
Roman law, 58
synodical, 34
Moral reckoning,
notion of, 87, 245
special meaning, 89

use, of,
available time, 239
Canon Law, 247
Roman law, 59

Natural reckoning,
definition of, 61
difficulty of application, 62, 209
occurrence, pre-Code, 79
Night, Roman division of, 39
Nones, 13
Noon, 42
Novitiate,
absence from,
import of the superior's permission, 206
practical duration of the day, 208
reckoning of the 30 days, 205-208
starting point of absence, 207
year of, question of the leap year, 192
Nundinae, 36

Obliquity of the ecliptic, 45
Old Style and New Style, 22
Olympiads, 26
Orders,
reckoning of the intervals, 198
suspension for illicit conferring of, 215
Ortus solis, 73

Parum pro nihilo reputatur,
applied to vacations, 128
Pasch, 55
Perihelion, 32
Physical reckoning, 88
Praescriptio longi temporis, 85
Precepts,
concurrence of, 165
of a private nature, pre-Code, 100
negative, 172
positive, 171
Prescription,
special reckoning, 116
thirty years, 85
Prolongation of time, 124
Public clocks and usual time, 144

Quamprimum, meaning of, 104

Reckoning,
actual aind intended use, 175
change of, 176, 179
choice of, 175
constituent factor, 175
Reckonings, various, for the same day,
constant and universal teaching, 168
different opinions, 159-162
lawfulness of, 163 ff.
not contradictory, 165
not restricted by Code, 164
possible scandal, 168
strange consequences, 167
successive days, 179
Recourse,
administrative processes, 243
dismissed religious, 243
rejection of libellus, 235
Regional time when public clocks indicate local time, 97
Restitutio in integrum,
Code, 243
Justinian, 69-70
Roman law, influence on Canon Law, 71
Rota, decisions on meaning of *usque*, 82

Seasons, Egyptian, 8
Servile works, 138
Ships, time followed by, 143
Sidereal time, use of in Canon Law, 152
Sirius, heliacal rising of, 8
Sothic cycle, 9

Spiritual exercises, 216
  starting point of, 205
Standard time,
  legality of when DST is imposed by law, 149, 151
    Europe, 149
    United States, 150
  shifting of zone lines, 151
  State laws, 151
Starting point,
  assignment of,
    explicit, 201
    implicit,
      instances in the Code, 204
      nature of, 203
      various opinions, 202-203
  coincidence with midnight,
    juridical, 219
    physical, 219
  knowledge of fact or fact itself, 233
Statim, duration of, 104
Sunrise, 41
Sunset, 41
Sunset to sunset reckoning, 91, 92
Suspensions,
  censures, 214
  comparison with appeals, 216
  nature of starting point, 214-216
  practical norm, 217
  vindictive penalties,
    condemnatory sentence, 216
    incurred *ipso facto,* 215
  when duration is not determined by law, 216

Tempus continuum de iure tantum, 214
Tempus utile, cf. available time
Tempus utile ratione initii,
  admissibility of in Canon Law, 231
Terminology,
  looseness of, 72, 97, 98
Terminus ad quem,
  pre-Code, 84
  Roman law, 63-64
Terminus a quo,
  pre-Code, 87
  Roman law, 63
Time,
  apparent sun time, 45
  definition of, 1
  equation of, 46
  importance of,
    in Canon Law, 3
    in general, 2
  legal, 145
    extraordinary, 148
    regional, 145-146
      introduction of in the United States, 47
  local, 143
  material reckoning of, 5
  mean solar, 43, 143
  measuring factors of, 5
  non-local, decision of the Holy See, 94
  reckoning opposed to use of, 247
  sidereal, 152
  Standard, 47-48
  true sun time, 43, 143
    use of, pre-Code, 98
  usual, 143
  zonal, 146
Time-reckoning,
  fundamental norm, 111,
  influence on,
    Jewish, 51
    Roman, 52
  Jewish, 52
  location of canons, 110
  validity of acts, 3
Time-zone map of the United States, 147
Travelers,
  International Date Line, 186
    repetition of a day, 186
    suppression of a day, 187
  zone to zone, 180

divine office, 184
fast and abstinence, 181
successive ember days, 182
Mass and Holy Communion, 185
eucharistic fast, 182
Trials, juridical,
duration of, 246

Unavailing days, 234
Urgere,
meaning of the word in contracts, 120-121
Usucapio, 84

Vacations,
question of the continuous reckoning, 212
reckoning of,
bishops, 126
canons, 126
starting point not determined, 205
Vernal equinox,
variation throughout the ages, 21
Vespers, starting point for the reckoning of the day, 114
Vicar Capitular,
election of, 232
Vows,
end of temporary vows, 228
renewal of, 224

Week,
arbitrary unit, 37
names of the days, 36-37
origin of, 35
reckoning of, 129
Roman law, 59
various kind of, 35
with the Jews, 55

Year,
anomalistic, 32
canonical, duration of, 73, 130
civil,
beginning of, 23 ff.
Jewish, 53
Roman, 11
legal, Roman, 59
sidereal, 31
tropical, 32
Years, regnal, 28

Zonal time,
decisions of the Holy See, 98-99
legality of, 146
usual time, 148
Zones, time,
division of the world into, 146

CANON LAW STUDIES

1. Freriks, Rev. Celestine A., C.PP.S., J.C.D., Religious Congregations in Their External Relations, 121 pp., 1916.
2. Galliher, Rev. Daniel M., O.P., J.C.D., Canonical Elections, 117 pp., 1917.
3. Borkowski, Rev. Aurelius L., O.F.M., J.C.D., De Confraternitatibus Ecclesiasticis, 136 pp., 1918.
4. Castillo, Rev. Cayo, J.C.D., Disertacion Historico-Canonica sobre la Potestad del Cabildo en Sede Vacante o Impedida del Vicario Capitular, 99 pp., 1919 (1918).
5. Kubelbeck, Rev. William J., S.T.B., J.C.D., The Sacred Pentitentiaria and Its Relations to Faculties of Ordinaries and Priests, 129 pp., 1918.
6. Petrovits, Rev. Joseph J.C., S.T.D., J.C.D., The New Church Law On Matrimony, X-461 pp., 1919.
7. Hickey, Rev. John J., S.T.B., J.C.D., Irregularities and Simple Impediments in the New Code of Canon Law, 100 pp., 120.
8. Klekotka, Rev. Peter J., S.T.B., J.C.D., Diocesan Consultors, 179 pp., 1920.
9. Wanenmacher, Rev. Francis, J.C.D., The Evidence in Ecclesiastical Procedure Affecting the Marriage Bond, 1920 (Printed 1935).
10. Golden, Rev. Henry Francis, J.C.D., Parochial Benefices in the New Code, IV-119 pp., 1921 (Printed 1925).
11. Koudelka, Rev. Charles J., J.C.D., Pastors, Their Rights and Duties According to the New Code of Canon Law, 211 pp., 1921.
12. Melo, Rev. Antonius, O.F.M., J.C.D., De Exemptione Regularium, X-188 pp., 1921.
13. Schaaf, Rev. Valentine Theodore, O.F.M., S.T.B., J.C.D., The Cloister, X-180 pp., 1921.
14. Burke, Rev. Thomas Joseph, S.T.D., J.C.D., Competence in Ecclesiastical Tribunals, IV-117 pp., 1922.
15. Leech, Rev. George Leo, J.C.D., A Comparative Study of the Constitution, "Apostolicae Sedis" and the "Codex Juris Canonici," 179 pp., 1922.
16. Motry, Rev. Hubert Louis, S.T.D., J.C.D., Diocesan Faculties According to the Code of Canon Law, II-167 pp., 1922.
17. Murphy, Rev. George Lawrence, J.C.D., Delinquencies and Penalties in the Administration and Reception of the Sacraments, IV-121 pp., 1923.
18. O'Reilly, Rev. John Anthony, S.T.B., J.C.D., Ecclesiastical Sepulture in the New Code of Canon Law, II-129 pp., 1923.

19. Michalicka, Rev. Wenceslas Cyrill, O.S.B., J.C.D., Judicial Procedure in Dismissal of Clerical Exempt Religious, 107 pp., 1923.
20. Dargin, Rev. Edward Vincent, S.T.B., J.C.D., Reserved Cases According to the Code of Canon Law, IV-103, pp., 1924.
21. Godfrey, Rev. John A., S.T.B., J.C.D., The Right of Patronage According to the Code of Canon Law, 153 pp., 1924.
22. Hagedorn, Rev. Francis Edward, J.C.D., General Legislation on Indulgences, II-154 pp., 1924.
23. King, Rev. James Ignatius, J.C.D., The Administration of the Sacraments to Dying Non-Catholics, V-141 pp., 1924.
24. Winslow, Rev. Francis Joseph, A.F.M., J.C.D., Vicars and Prefects Apostolic, IV-149 pp., 1924.
25. Correa, Rev. Jose Servelion, S.T.L., J.C.D., La Potestad Legislativa de la Iglesia Catolica, IV-127 pp., 1925.
26. Dugan, Rev. Henry Francis, A.M., J.C.D., The Judiciary Department of the Diocesan Curia, 87 pp., 1925.
27. Keller, Rev. Charles Frederick, S.T.B., J.C.D., Mass Stipends, 167 pp., 1925.
28. Paschang, Rev. John Linus, J.C.D., The Sacramentals According to the Code of Canon Law, 129 pp., 1925.
29. Pointek, Rev. Cyrillus, O.F.M., S.T.B., J.C.D., De Indulto Exclaustrationis necnon Saecularizationis, XIII-289 pp., 1925.
30. Kearney, Rev. Richard Joseph, S.T.B., J.C.D., Sponsors at Baptism According to the Code of Canon Law, IV-127 pp., 1925.
31. Bartlett, Rev. Chester Joseph, A.M., LL.B., J.C.D., The Tenure of Parochial Property in the United States of America, V-108 pp., 1926.
32. Kilker, Rev. Adrian Jerome, J.C.D., Extreme Unction, V-425 pp., 1926.
33. McCormick, Rev. Robert Emmett, J.C.D., Confessors of Religious. VIII-266 pp., 1926.
34. Miller, Rev. Newton Thomas, J.C.D., Founded Masses According to the Code of Canon Law, VII-93 pp., 1926.
35. Roelker, Rev. Edward G., S.T.D., J.C.D., Principles of Privilege According to the Code of Canon Law, XI-166 pp., 1926.
36. Bakalarczyk, Rev. Richardus, M.I.C., J.U.D., De Novitiatu, VIII-208 pp., 1927.
37. Pizzuti, Rev. Lawrence, O.F.M., J.U.L., De Parochis Religiosis, 1927. (Not printed).
38. Bliley, Rev. Nicholas Martin, O.S.B., J.C.D., Altars According to the Code of Canon Law. XIX-132 pp., 1927.
39. Brown, Mr. Brendan Francis, A.B. LL.M., J.U.D., The Canonical Juristic Personality with Special Reference to Its Status in the United States of America. V-212 pp., 1927.

40. Cavanaugh, Rev. William Thomas, C.P., J.U.D., The Reservation of the Blessed Sacrament, VIII-101 pp., 1927.
41. Doheny, Rev. William J., C.S.C., A.B., J.U.D., Church Property: Modes of Acquisition, X-118 pp., 1927.
42. Feldhaus, Rev. Aloysius H., C.PP.S., J.C.D., Oratories, IX-141 pp., 1927.
43. Kelly, Rev. James Patrick, A.B., J.C.D., The Jurisdiction of the Simple Confessor, X-208 pp., 1927.
44. Neuberger, Rev. Nicholas J., J.C.D., Canon 6 or the Relation of the Codex Juris Canonici to the Preceding Legislation, V-95 pp., 1927.
45. O'Keefe, Rev. Gerald Michael, J.C.D., Matrimonial Dispensations, Powers of Bishops, Priests and Confessors, VIII-232 pp., 1927.
46. Quigley, Rev. Joseph A.M., A.B., J.C.B., Condemned Societies, 139 pp., 1927.
47. Zaplotnik, Rev. Johannes Leo, J.C.D., De Vicariis Foraneis, X-142 pp., 1927.
48. Duskie, Rev. John Aloysius, A.B., J.C.D., The Canonical Status of the Orientals in the United States, VIII-196 pp., 1928.
49. Hyland, Rev. Francis Edward, J.C.D., Excommunication, Its Nature, Historical Development and Effects, VIII-181 pp., 1928.
50. Reinmann, Rev. Gerald Joseph, O.M.C., J.C.D., The Third Order Secular of Saint Francis, 201 pp., 1928.
51. Schenk, Rev. Francis J., J.C.D., The Matrimonial Impediments of Mixed Religion and Disparity of Cult, XVI-318 pp., 1929.
52. Coady, Rev. John Joseph, S.T.D., J.U.D., A.M., The Appointment of Pastors, VIII-150 pp., 1929.
53. Kay, Rev. Thomas Henry, J.C.D., Competence in Matrimonial Procedure, VIII-164 pp., 1929.
54. Turner, Rev. Sidney Joseph, C.P., J.U.D., The Vow of Poverty, XLIX-217 pp., 1929.
55. Kearney, Rev. Raymond, A., A.B., S.T.D., J.C.D., The Principles of Delegation, VII-149 pp., 1929.
56. Conran, Rev. Edward James, A.B., J.C.D., The Interdict, V-163 pp., 1930.
57. O'Neil, Rev. William H., J.C.D., Papal Rescripts of Favor, VII-218 pp., 1930.
58. Bastnagel, Rev. Clement Vincent, J.U.D., The Appointment of Parochial Adjutants and Assistants, XV-257 pp., 1930.
59. Ferry, Rev. William A., A.B., J.C.D., Stole Fees, V-135 pp., 1930.
60. Costello, Rev. John Michael, A.B., J.C.D., Domicile and Quasi-domicile, VII-201 pp., 1930.
61. Kremer, Rev. Michael Nicholas, A.B., S.T.B., J.C.D., Church Support in the United States, VI-1930.

62. Angulo, Rev. Luis, C.M., J.C.D., Legislation de la Iglesia sobre la intencion en la application de la Santa Misa, VII-104 pp., 1931.
63. Frey, Rev. Wolfgang Norbert, O.S.B., A.B., J.C.D., The Act of Religious Profession, VIII-174 pp., 1931.
64. Roberts, Rev. James Brendan, A.B., J.C.D., The Banns of Marriage, XIV-140 pp., 1931.
65. Ryder, Rev. Raymond Aloysius, A.B., J.C.D., Simony, IX-151 pp., 1931.
66. Campagna, Rev. Angelo, Ph.D., J.U.D., Il Vicario Generale del Vescovo, VII-205 pp., 1931.
67. Cox, Rev. Joseph Godfrey, A.B., J.C.D., The Administration of Seminaries, VI-124 pp., 1931.
68. Gregory, Rev. Donald J., J.U.D., The Pauline Privilege, XV-165 pp., 1931.
69. Donohue, Rev. John F., J.C.D., The Impediment of Crime, VII-110 pp., 1931.
70. Dooley, Rev. Eugene A., O.M.I., J.C.D., Church Law On Sacred Relics, IX-143 pp., 1931.
71. Orth, Rev. Raymond Clement, O.M.C., J.C.D., The Approbation of Religious Institutes, 171 pp., 1931.
72. Pernicone, Rev. Joseph M., A.B., J.C.D., The Ecclesiastical Prohibition of Books, XII-267 pp., 1932.
73. Clinton, Rev. Connell, A.B., J.C.D., The Paschal Precept, IX-108 pp., 1932.
74. Donnelly, Rev. Francis B., A.M., S.T.L., J.C.D., The Diocesan Synod, VIII-125 pp., 1932.
75. Torrente, Rev. Camilo, C.M.F., J.C.D., Las Processiones Sagradas, V-145 pp., 1932.
76. Murphy, Rev. Edwin J., C.PP.S., J.C.D., Suspension Ex Informata Conscientia, XI-122, pp., 1932.
77. Mackenzie, Rev. Eric F., A.M., S.T.L., J.C.D., The Delict of Heresy in its Commission Penalization, Absolution, VII-124 pp., 1932.
78. Lyons Rev. Avitus E., S.T.B., J.C.D., The Collegiate Tribunal of First Instance, XI-147 pp., 1932.
79. Connolly, Rev. Thomas A., J.C.D., Appeals, XI-195 pp., 1932.
80. Sangmeister, Rev. Joseph V., A.B., J.C.D., Force and Fear as Precluding Matrimonial Consent, V-211 pp., 1932.
81. Jaeger, Rev. Leo A., A.B., J.C.D., The Administration of Vacant and Quasi-vacant Episcopal Sees in the United States, IX-229 pp., 1932.
82. Rimlinger, Rev. Herbert T., J.C.D., Error Invalidating Matrimonial Consent, VII-79 pp., 1932.
83. Barrett, Rev. John D.M., S.S., J.C.D., A Comparative Study of the Third Plenary Council of Baltimore and the Code, IX-221 pp., 1932.

84. Carberry, Rev. John J., Ph.D., S.T.D., J.C.D., The Juridical Form of Marriage, X-177 pp., 1934.
85. Dolan, Rev. John L., A.B., J.C.D., The Defensor Vinculi, XII-157 pp., 1934.
86. Hannan, Rev. Jerome D., A.M., S.T.D., LL.B., J.C.D., The Canon Law of Wills, IX-517 pp., 1934.
87. Lemieux, Rev. Delisle A., A.M., J.C.D., The Sentence in Ecclesiastical Procedure, IX-131 pp., 1934.
88. O'Rourke, Rev. James J., A.B., J.C.D., Parish Registers, VII-109 pp., 1934.
89. Timlin, Rev. Bartholomew, O.F.M., A.M., J.C.D., Conditional Matrimonial Consent, X-381 pp., 1934.
90. Wahl, Rev. Francis X., A.B., J.C.D., The Matrimonial Impediments of Consanguinity and Affinity, VI-125 pp., 1934.
91. White, Rev. Robert J., A.B., LL.B., S.T.B., J.C.D., Canonical Ante-Nuptial Promises and the Civil Law, VI-152 pp., 1934.
92. Herrera, Rev. Antonio Parra, O.C.D., J.C.D., Legislation Ecclesiastica sobra el Ayuno y la Abstinencia, XI-191 pp., 1935.
93. Kennedy, Rev. Edwin J., J.C.D., The Special Matrimonial Process in Cases of Evident Nullity, X-165 pp., 1935.
94. Manning, Rev. John J., A.B., J.C.D., Presumption of Law in Matrimonial Procedure, XI-111 pp., 1935.
95. Moeder, Rev. John M., J.C.D., The Proper Bishop for Ordination and Dismissorial Letters, VII-135 pp., 1935.
96. O'Mara, Rev. William A., A.B., J.C.D., Canonical Causes For Matrimonial Dispensations, IX-155 pp., 1935.
97. Reilly, Rev. Peter, J.C.D., Residence of Pastors, IX-81 pp., 1935.
98. Smith, Rev. Mariner T., O.P., S.T.L., J.C.D., The Penal Law For Religious, VII-169 pp., 1935.
99. Whalen, Rev. Donald W., A.M., J.C.D., The Value of Testimonial Evidence in Matrimonial Procedure, XIII-297 pp., 1935.
100. Cleary, Rev. Joseph F., J.C.D., Canonical Limitations on the Alienation of Church Property, VIII-141 pp., 1936.
101. Glynn, Rev. John C., J.C.D., The Promoter of Justice, XX-337 pp., 1936.
102. Brennan, Rev. James H., S.S., A.M., S.T.B., J.C.D., The Simple Convalidation of Marriage, VI-135 pp, 1937.
103. Brunini, Rev. Joseph Bernard, J.C.D., The Clerical Obligations of Canons, 139 and 142, X-121 pp., 1937.
104. Connor, Rev. Maurice, A.B., J.C.D., The Administrative Removal of Pastors, VIII-159 pp., 1937.
105. Guilfoyle, Rev. Merlin Joseph, J.C.D., Custom, XI-144 pp., 1937.
106. Hughes, Rev. James Austin, A.B., A.M., J.C.D., Witnesses in Criminal Trials of Clerics, IX-140 pp., 1937.

107. Jansen, Rev. Raymond J., A.B., S.T.L., J.C.D., Canonical Provisions for Catechetical Instruction, VII-153 pp., 1937.
108. Kealy, Rev. John James, A.B., J.C.D,, The Introductory Libellus in Church Court Procedure, XI-121 pp., 1937.
109. McManus, Rev. James Edward, C.SS.R., J.C.D., The Administration of Temporal Goods in Religious Institutes, XVI-196 pp., 1937.
110. Moriarity, Rev. Eugene James, J.C.D., Oaths in Ecclesiastical Courts, X-115 pp., 1937.
111. Rainer, Rev. Eligius George, C.SS.R., J.C.D., Suspension of Clerics, XVII-249 pp., 1937.
112. Reilly, Rev. Thomas F., C.SS.R., J.C.D., Visitation of Religious, VI-195 pp., 1938.
113. Moriarty, Rev. Francis E., C.SS.R., J.C.D., The Extraordinary Absolution from Censures, XV-334 pp., 1938.
114. Connolly, Rev. Nicholas P., J.C.D., The Canonical Erection of Parishes, X-132 pp., 1938.
115. Donovan, Rev. James Joseph, J.C.D., The Pastor's Obligation in Prenuptial Investigation, XII-322 pp., 1938.
116. Harrigan, Rev. Robert J., M.A., S.T.B., J.C.D., The Radical Sanation of Invalid Marriages, VIII-208 pp., 1938.
117. Boffa, Rev. Conrad Humbert, J.C.D., Canonical Provisions for Catholic Schools, X-211 pp., 1939.
118. Parsons, Rev. Anscar John, O.M. Cap., J.C.D., Canonical Elections. XII-236 pp., 1939.
119. Reilly, Rev. Edward Michael, A.B., J.C.D., The General Norms of Dispensation, X-156 pp., 1939.
120. Ryan, Rev. Gerald Aloysius, A.B., J.C.D., Principles of Episcopal Jurisdiction, XII-172 pp., 1939.
121. Burton, Rev. Francis James, C.S.C., A.B., J.C.D., A Commentary on Canon 1125, X-222 pp., 1940.
122. Miaskiewicz, Rev. Francis Sigismund, J.C.D., Supplied Jurisdiction according to Canon 209, XII-340 pp., 1940.
123. Rice, Rev. Patrick William, A.B., J.C.D., Proof of Death in Prenuptial Investigation, VIII-156 pp., 1940.
124. Anglin, Rev. Thomas Francis. M.S., J.C.L., The Eucharistic Fast.
125. Coleman, Rev. John Jerome, J.C.L., The Minister of Confirmation.
126. Downs, Rev. John Emmanuel, A.B., J.C.L., The Concept of Clerical Immunity.
127. Esswein, Rev. Anthony Albert, J.C.L., Extrajudicial Penal Powers of Ecclesiastical Superiors.
128. Farrell, Rev. Benjamin Francis, M.A., S.T.L., J.C.L., The Rights and Duties of the Local Ordinary Regarding Congregations of Women Religious of Pontifical Approval.

129. Feeney, Rev. Thomas John, A.B., S.T.L., J.C.L., Restitutio in Integrum.
130. Findlay, Rev. Stephen William, O.S.B., A.B., J.C.L., Canonical Norms Governing the Deposition and Degradation of Clerics.
131. Goodwine, Rev. John, A.B., S.T.L., J.C.L., The Right of the Church to Acquire Property.
132. Heston, Rev. Edward Louis, C.S.C., Ph.D., S.T.D., J.C.L., The Alienation of Church Property in the United States .
133. Hogan, Rev. James John, S.T.L.; J.C.L.,, Judicial Advocates and Procurators.
134. Kealy, Rev. Thomas M., A.B., Litt. B., J.C.L., Dowry of Women Religious.
135. Keene, Rev. Michael James, O.S.B., J.C.L., Religious Ordinaries and Canon 198.
136. Kerin, Rev. Charles A., S.S., M.A., S.T.B., J.C.L., The Privation of Christian Burial.
137. Louis, Rev. William Francis, M.A., J.C.L., Diocesan Archives.
138. McDevitt, Rev. Gilbert Joseph, A.B., J.C.L., Legitimacy and Legitimation.
139. McDonough, Rev. Thomas Joseph, A.B., J.C.L., Apostolic Administrators.
140. Meier, Rev. Carl Anthony, A.B., J.C.L., Penal Administrative Procedure Against Negligent Pastors.
141. Schmidt, Rev. John Rogg, A.B., J.C.L., The Principles of Authentic Interpretation in Canon 17 of the Code of Canon Law.
142. Slafkosky, Rev. Andrew Leonard, A.B., J.C.L., The Canonical Episcopal Visitation of the Diocese.
143. Swoboda, Rev. Innocent Robert, O.F.M., J.C.L., Ignorance in Relation to the Imputability of Delicts.
144. Dubé, Rev. Arthur Joseph, A.B., J.C.L., The General Principles for the Reckoning of Time in Canon Law.
145. McBride, Rev. James T., A.B., J.C.L., Incardination and Excardination of Seculars.

www.ingramcontent.com/pod-product-compliance
Lightning Source LLC
LaVergne TN
LVHW050257080826
844660LV00012B/649

* 9 7 8 0 8 1 3 2 2 3 3 3 9 *